Daqna Gabitkyzy Ospanowa

Women's Leadership in Central Asia: Strategy and Priorities

Daqna Gabitkyzy Ospanowa

Women's Leadership in Central Asia: Strategy and Priorities

ScienciaScripts

Cover image: www.ingimage.com

This book is a translation from the original published under ISBN 978-620-4-18421-0.

Publisher:
Sciencia Scripts
is a trademark of
Dodo Books Indian Ocean Ltd., member of the OmniScriptum S.R.L Publishing group
str. A.Russo 15, of. 61, Chisinau-2068, Republic of Moldova Europe
Printed at: see last page
ISBN: 978-620-4-08901-0

CONTENTS

INTRODUCTION.. 3

1 THE ROLE AND STATUS OF WOMEN IN THE HISTORY OF THE PEOPLES OF CENTRAL ASIA 7

2. POLITICAL LEADERSHIP IN CENTRAL ASIAN COUNTRIES... 29

CONCLUSION.. 47

LIST OF REFERENCES USED: 50

LIST OF ABBREVIATIONS AND ACRONYMS

ABRA Asian Development Bank
CEDAW Convention on the Elimination of All Forms of Discrimination against Women
CEDAW- OP Optional Protocol to the Convention on the Elimination of All Forms of Discrimination against Women
FAO Food and Agriculture Organization of the United Nations
IFCInternational Federation of Surveyors
GDPGross Domestic Product
GIRGENDER and development
HDI Human Development Index
ICCPR-International Covenant on Civil and Political Rights
ICESCR International Covenant on Economic, Social and Cultural Rights
IFAD International Fund for Agricultural Development
ILO International Labour Organization
UNDP NGO/UNDP OOH Development Programme
UNECE-European Economic Commission OOH
UNRISD Research Institute for Social Development at OOH

INTRODUCTION

This thesis is devoted to the study of the gender situation in the Central Asian region and its possible perspectives.

The relevance of the research lies in the necessity to analyze the gender situation in modern Central Asian states, as well as to compare with the societies of traditional Turkic tribes and the societies of these states during the Soviet period. It is also important to at the same time is important: study of the strategies and mechanisms of each state in Central Asia to improve gender equality. Gender inequality is root cause many obstacles sustainable evelopment and the complexity of the the road to democratization. Despite the efforts of all states in the region to liberalize all spheres of life, gender inequalities are still present not only in social life, but also in other spheres of society. Obviously, no sphere of life can be gender-neutral, but the gender divide has a destructive impact on such sectors as the economy, political and administrative activities, culture, and so on. A modern state can make progress or achieve high rates of development only when both men and women have the right to equal opportunities. The countries of Central Asia have experienced significant socio-economic,cultural and political changes accompanied by major social changes,gradual economicgrowth with gradual economic growth and a slight improvement in living standards. The region as a whole has also seen an end to rampant gender inequalities. genderglaring gender gaps have largely been eliminated,However, large inequalities persist and disproportionately affect the lives of girls and women. Women in Central Asia are often subjected to patriarchal pressures to attain equal rights to health, education, decision-making and economic independence in terms of wages. Some of the aspects of gender inequality are deeply entrenched patriarchal norms, the low value of girls and persistent gender stereotypes and prejudices that are difficult to overcome in the region. Progress has been slow in women's participation and representation in political and economic life in Central Asia. Women have generally not been able to command more authority in government and rarely lead sectoral ministries, senior government bodies and major political parties. Occupational segregation of women, concentrated in low-paying informal sectors of social security, persists and is perpetuated through gender-based exclusion stereotypes about the types of pa6o tics, which are divided into typically "male" or "female". In order to eradicate this problem, many nongovernmental organizations on various topics and interests are functioning in Central Asian countries, promoting women in governmental structures. Basically, these NGOs are based on sound experience. To date, their results and progress cannot be compared with similar organizations working in European countries, but they have a positive impact on the achievement of indicators in the countries on this issue.

Degree of study of the topic.

Women's leadership in Central Asia has not yet been sufficiently studied, as it is in the initial stages of development. Three types of sources were used in this dissertation.

This includes monographs, scientific articles and other papers by Western and national authors that include gender policies in Central Asian countries, information and current data on the situation of women, as well as recommendations for improving the gender situation in the region.

Most of the research and analysis in this area is carried out by domestic and international organizations.

In the process of writing the diploma work were used by such Russian authors as Ykkem6aeva M.A., Shegenova J.H., Burkhanov K.H. H., Baisakova Z.M., Ishmukhamedova L.I., Kaltaeva L.M., Ko6eeva A.O., Shakirova C. M., Beke6aeva A. D and so on.

Ykkem6aeva M. A. in her research describes working conditions for women, possible risks and losses for women in the labor market, as well as principles of gender analysis of trade policy. The author notes that women in the labor market, being self-employed, do not receive a social package, witness unsatisfactory working conditions and often do not receive poco6ya and pension contributions [1].

The main topic of Schegenova's work is the history and development of gender policy in Kazakhstan, as well as the main directions, strategies, mechanisms and initiatives of the government. The author noted that the first major changes in gender policies were undertaken in the context of the crisis of the 1990s. To date, gender policy in Kazakhstan is one of the methods of formation of the post-crisis era [2].

A.D. Beke6aeva's article studies women's non-governmental organizations in Central Asia [3]. Burkhanov K. H., Baisakova Z.M., Ishmukhamedova L.I., Kaltaeva L.M., Ko6eeva A.O., Shakirova C. M in their work reveal the role and work of the women's movement in Kazakhstan as an aspect of social modernization. This paper also explores such aspects as topical issues women's movement, functioning of crisis centers, risks and opportunities, thematic NGOs and organizations [4].

Lipovka A. highlighted the topic of gender segregation of labour in higher education in Kazakhstan, noting that gender segregation leads to a decrease in gender capacity, which ultimately leads to mass dissatisfaction of Kazakhstanis with the system of higher education in the state.

This includes important international declarations and treaties on women's lives everywhere, as well as mechanisms and strategies to enhance the role of women in economic, political, cultural, social and other spheres. They include the 1995 Beijing Declaration and Platform for Action, the 1997 United Nations Convention on the Elimination of All Forms of Discrimination against Women, the 1993 United Nations Declaration on the Elimination of Violence against Women, the 1993 Vienna Declaration and Programme of Action and the 2000 Political Declaration on Women 2000: Gender Equality, Development and Peace for the Twenty-first Century.

This includes information from news portals, governmental Internet resources and the like.

The object of this thesis is the role of women in different spheres of life in Central Asia. The role of women in different spheres of life in Central Asian countries is the object of study of this thesis.

The subject of this thesis. The subject of the research is mechanisms to overcome gender discrimination in society, as well as methods to eliminate the influence of patriarchal conservative ideas regarding the status of women in Central Asian countries.

Objective of the thesis. To analyse each state's strategies on women's issues and examine the actual outcomes.

In order to achieve this goal, the following tasks were formed:

1. Examine the basic documents relating to women's rights in each state separately;
2. Compare the current position of women with other time periods in the Central Asian region;
3. Aanalizirovatioonn oo womeen in the o6present life of the state;
4. Identify key issues and problems that constitute a barrier to gender equality in the region;
5. Izychytkotnopravlenie kontsentratsiynnogocyadpravtvavljaet women's status in the state;
6. To make women active participants in the development and establishment of the State;
7. Examine the level of o6paΘsΘcation of women;
8. Examine the gender institutions that operate in each country;
9. Identify the main ways to address the women's issue in the Central Asian region.

Theoretical and methodological basis of the thesis. The theoretical basis of the research is the works of Western and domestic analysts, a number of methodological frameworks dedicated to the main problems of women in the region of Central Asia. The following methods were used: system analysis, chronological analysis, comparative analysis and historical and political analysis.

Chronological framework of the diploma work. The chronological framework of the diploma work covers the period from the 16th century to the present.

The **structure of the diploma work.** The diploma paper consists of an introduction, two chapters, six paragraphs, conclusion, list of references.

The introduction will outline the relevance of the diploma thesis, describe its main goals and objectives, its purpose, theoretical and practical aspects, as well as the structure of the thesis.

The first chapter examines the situation of women in the Central Asian region at different points in time in order to compare living conditions, rights and opportunities with the present time.

The second chapter describes current data on the participation of Central Asian women in public and political life in the region, as well as legal and policy frameworks, information on assistance from international organizations, and government results over the last three decades.

Conclusion contains the main conclusions and results of the study.

The **practical significance of the work** lies in the possibility of using the results of the study to solve practical problems in this field.

1. THE ROLE AND STATUS OF WOMEN IN THE HISTORY OF THE PEOPLES OF CENTRAL ASIA

1.1 The position of women in traditional societies of Central Asia

The social and cultural life of the ancient Turks has always been an interesting aspect for historians to study, because it is very different from other cultures and has its own unique oco6eonnocities. Since the woman is one of the 6aZoVa figures in the ancient Turkic chronicles, the study of the female issue is also quite profound and has o6pensive material and written 6aZo.
In the study of Turkic history, we can conclude that women have always had a significant role not only as a nucleus of society, but also had special rights in political and military life. The gender issue among the ancient Turks is determined by two sources: writings and archaeological artifacts. The lion's share of information about their social and cultural life dates from the 6th-10th centuries. A significant role in this question is played by the ancient pogpe6e6ons, by which we can determine not only the financial state of the buried, but also their "level" among the people. Thus, if traditionally the man occupied a dominant position in all respects, namely as a warrior, breadwinner and provider, a lot of material and written evidence informs differently. In particular, the funeral complexes of the Scythian and Sarmatian period, notably the tombs of prominent women, indicate with certainty that female capability was recognized among the tribes of that time. The women's graves and the objects buried with them look much more beautiful than those of the men, as historians have also shown, when studying the archaeological excavations of the ancient Turkic plains. The above said shows that in comparison with the life of women in the ancient Chinese, the female sex of this nation was of much higher status.Initially, the status of women in the society of ancient tribes inhabiting the territory of present-day Central Asia was defined as"patroness of the home". That is, since the pastoral form of life is nomadic, this formulated the further o6язanocity of women in o6ecculture. The main activities of men are hunting, providing food, means of survival for their families, while the main activities of women are creating necessary conditions for subsistence, raising children and storing and preparing food for the family. This is confirmed by the line from the book "Kita6-i Dadem Korkut", which is the only surviving book of the epos describing the life of Turkic Oguzes, that an ideal wife is someone who, "Supports her home, is someone who, when a guest comes to the house from the steppe, when her husband is out hunting, she feeds that guest, gives him drink, respects him and lets him go". The husband should fully protect and provide for

his wife. "The bright-eyed brides "Yours, yours - not mine! But there is one there that is given to me, Whether it be to forty or one foe, I will never give it up! - The lines from "Kita6-i Dedem Korkut" [1] read.

In Turkic families, kinship was defined not only by the father, but also by the mother. This is confirmed by the fact that the mother and the father had the same status. If the husband had the status of "lord of the house", his wife was called "mistress of the house." A person whose mother was not of Turkic origin was not considered a pure-blooded person. Turkic women, like men, had the right to land tenure and a co6cve income. When the male soldiers marched on foot, women were transported together with their children on wagons. The rape of a woman by the Turks was characterized as a grave crime, on a par with such crimes as genocide, murder, etc., and, accordingly, was punished by the death penalty [2].

At the same time, Turkic women occupied not the last niche in the development of political and social life, and their activities included not only housekeeping and childbirth. For example, a very important strategic method of the Turks were the so called "dynastic" unions, which had a strong influence on both the possible splits and the prosperity of the state. In the history there is evidence of a number of feudal 6paoks concluded by the Turkic states with Chinese Tang dynasty, where the latter had the purpose to spread their influence in the neighboring countries. Also, in many Turkic tribes, women could actively participate in decision of political vital issues, be present at tojas, kurultais, where warriors and the wisest members of the Society were co6ipresented. At the same time, the Turkic khagan always consulted with his wife before making a concrete decision. If his wife rejected his decision, it was found invalid and never came into force. The official historical chronicle of the Chinese Tang dynasty Jiu Tang Shu, known as the "Old Tang history", tells that in the Turkic tribes living in the north for a long time there has been a tradition of o6ychai, where the "katun", i.e. woman, took part in military campaigns. They could also occupy the positions of ambassadors, commanders and warlords. Turkic women were also skilled in the art of war, namely riding, archery and hunting on a par with the male warriors. Despite the militancy and strength of Turkic women, men valued them for their chastity, wisdom, and femininity [3]. For example, the apa6c historian and writer A6y Ysman 6n Al Djahiz in his numerous monographs about Turks describes women in such a way: "Their women are made in accordance with the o6pa6sion and likeness of their men, and their horses are made exclusively for them". The Turkic bride was not only the ideal housewife, she was also her man's support in military affairs. According to the source."In Kita6-i Daedem Korkut, there is a description of a moment in the book where an elderly father asks his son, who has come of age, whatThe son answered: "Father, while I have not yet risen from my seat, let her rise; while I have not yet mounted my steed, let her mount; while I have not yet gone to the people of the bloodthirsty gyaur, let her go, let her bring me the head". In this book, three women - Burla Khatun, Selchen Khatun and Banu Chichek are described as female warriors, with excellent physical training, who went together with men to fight against enemies. Burla Khatun in "Kita6-i Daedem Korkut" is described as follows: "She put forty slender maidens on horses, ordered to bring a raven-haired animal, sat on horseback, girded herself with a sword and went [to search for her son]... Madam Burla the Tall was a black banner of gyaur with her sword and trampled it into the ground" [4].

As is peculiar to the Eastern culture, the respect to women of the Turks occupied a very important and sacral place in their worldview. A woman was considered the progenitor of the family. In the book Kita6-i Daedem Korkut, Khan addresses his mother: "Come, mother who gave birth to me, With her 6eel milk you nourish me, Mother, with your soul you have cop6eated me, Mother, who is gray with grief..." The Türks traditionally worshipped Yomay, mother of children and 6emale girls, the ancestress of the Türks, who could turn into a liar. As a worship Ymai was worshipped either by the mountains, caves, and the like, or by the Sun, or fire. Ymai goddess was always associated with the darkness, therefore the exploiting of nature, tearing the grass, and so on was condemned by the Türks. It was considered that if to spite Mai, it can deprive all living beings of the ability to fertilize. Besides that, the Turks had other highly esteemed female 6ojects: Ak Ana - Goddess of Creation, Yel Ana - Goddess of winds, Cy Ana - Goddess of water, Od Ana - Goddess of fire, Bayanai - Goddess of hunting, Kaylin - Goddess of kingdoms, Aisyt - Goddess of beauty, Pushka Ana - Goddess of the sun, Kuyash - Goddess of creation. In the territory of the present Central Asia were found small figures representing a woman, as 6o goddess of fertility. The Turks worshipped each of them and had a precept to sacrifice to them so as not to anger the gods and to be in harmony with them.

In the book "The New History of Tang dynasty" there is an episode in which the son of the Turkic kagan refused to start a feast because his mother did not get a portion of meat. This so struck the Chinese emperor that he ordered to immediately bring the woman her share of meat [5].

However, in the Turkic traditional o6щщtv not all women had the same rights and authority. For example, based on the sources of Turkic written monuments, the most authoritative woman wore the proud title "qatun". The only woman who had such a status was the mother of Bilge-Kagan. Along with the title, she was also a full-fledged heiress of her father and had equal rights to the throne.

Moreover, since in the Turkic culture polygyny became a common thing after In the early Middle Ages, the Muslims were the only women who had privileges over the others. It is believed that the lowest rung of the social ladder was occupied by concubines, captured foreign women (mostly from China and Iran). Their lives after being captured developed in different ways: they either became wives of Turkic nomads of the same origin, or they were noticed by wealthy members of the tribes who subsequently took them as wives. Traditionally, concubines were always less respected than Turkic wives, as the latter always had support in the form of relatives. Despite this fact, children of concubines and children of Turkic wives had the same rights, regardless of their origin [6].

As has already been written before, the most information about the position of the woman in the traditional Turkic culture is given by the log-people's complexes, because they fully provide a great deal of physical evidence. Up to now about 400 log complexes have been discovered and registered in the territory of Central Asia and they definitely belong to the Turkic culture. Only fewer than 200 graves have been studied thoroughly and in detail. The study of these graves was very difficult, there are several reasons for that: many graves were barbarically grazed, others were destroyed to such a degree that it was impossible to determine the gender of buried people. Obviously, if the skeleton of the buried person was heavily mutilated, it was impossible to determine the sex, and his inventory did not have a clear gender representation, archaeologists considered him to be female. Turkic burials, more specifically the interior, have a tendency toward standardization and eadin 6paization, which makes determining gender difficult. Thus, about o6pao

130 graves were recognized as male, about 40 as female, and the remaining graves were burials of children and adolescents. This ratio of male and female burial complexes is due to the fact that in the Middle Ages in Central Asia, due to the active reasons - wars and warfare, a shaky political system - the male mortality was many times higher than the female one. Moreover, this tendency is inherent in nomadic life styles. In the study of ancient burials, archaeologists agreed that women's graves were much more

The graves of the women were "more luxurious" and "more beautiful" than those of the men. Predominantly, women's graves contained objects "typical for the female sex": jewelry and precious metals (earrings, rings, rings, necklaces, necklaces), mirrors made of metal, remnants of natural silk clothes, yarn and tools for knitting and sewing. In some burial sites were found objects that do not belong to the traditionally female objects, i.e. weapons (axes, daggers, knives), equestrian equipment and fragments of military equipment. Thus

In this way, the people of the region were able to see that they were not the only ones in the world who had been buried there. Another explanation is that from ancient times the Türks believed in "married life", i.e. if a woman was widowed, and then subsequently died, her grave contained man's clothes, her husband's clothes, to be ready to meet him in the other world. The average age of buried women was estimated by archaeologists as 40-50 years old, which explains the reverence for adult women among the Turks [7].

It is not unimportant that the Kaghan complexes have always been accompanied by their stone sculptures of their wives. For example, together with the monument of Kul-Tegin the remains of his wife's sculpture, more precisely her head, were found. Archaeologists have reported that the female sculptures found were very detailed, especially the facial features, on which they noted "the facial expression was very strong and severe".

It follows from all the above that the woman in the traditional Turkic society occupied a respected high position, was a full-fledged and respected member of society, combining in her personas of a faithful and devoted wife, a good mother, a homemaker, and even a warrior, defender of her state.

1.2. Women's Leadership in the Years of Soviet Power

The 20th century is considered as a landmark in addressing the improvement of gender policies around the world, as it was the period when various reforms and trends for women's self-determination started. In the Soviet Union, the term"Gender" has been introduced in order to gradually eliminate discrimination of any kind on the basis of gender, as well as to focus on gender inequalities.Since the creation of the Soviet Union, legislation has declared legal equality between the sexes. However, as Lenin pointed out, there was no equality as such in all spheres of life. He called for equality under the slogan that managing, industrious women would "catch up" to men, and that women's hard work would lead to positive results and full socialism. B ideally, he believed that the o6ojective enterprises should employ a c6o mulation of the female population of the union pecpy6lics. [8]

Shortly after the 1917 6olshevik revolution, the Soviets took control of Central Asia, o6paoing five new pecpy6liks: Kazakhstan, Kyrgyzstan, Turkmenistan, Yz6eqiqtan and Tajikistan. Y establishment of Soviet power seriously damaged the traditional yep6 life in these pecpy6liks, which had been dominated for centuries by Muslim beliefs and o6 erities. In the absence of the indigenous proletariat as a 6aase of support for the new regime, he women-their activities and rights of the region were strictly limited by Sharia law and local traditions, which were 6o6pawned to play the role of "surrogate proletariat". The so-called emancipation of the women of Central Asia, whom Vladimir Lenin considered "the na6oee of the poppa6oed, the na6oee of the oppressed", became the central pillar of the Soviet effort in the region.

The road to maturation that Central Asian women have travelled has been long and arduous. They often faced resistance from their women and families. As elsewhere in the CCCP, women's new role as professionals and builders of communism did not fully supplant their traditional roles as daughters, wives and mothers, forcing them to bear the double burden of responsibility for their families.

Nevertheless, the reforms promoted by the Soviets gradually took root. The propaganda photographs in this exhibition, taken in the late 1940s, show a radically changed Central Asia. Although these photographs are artificially optimistic, sometimes obviously staged and often concealing the harsh realities of collectivization, they reflect a mood of genuine joie de vivre.

The "ocw6o6ed" successful women and girls seen here are real. It's the same with the optimism and self-esteem they radiate in the face of renewed equality and opportunity.

First, legislation was enacted to protect women's rights and equalize them with those of men in all areas of political, economic and cultural life. In 1921, the Government of Turkmenistan issued a decree setting the minimum age for women to enter into marriage at 16 years of age, compared to 9 years under Shariah law. It also prohibited polygamy, forced marriage and kalim (bride price). The new norms were to be promoted amongst the o6ecdentry through women's sections, also known as Women's Departments, established within the o6lact and city committees of the Communist Party after 1919. Women's Department activists were responsible for organizing women's literacy schools, orphanages, kindergartens, and child-rearing consultations. .

On 8 March 1927 activists in Tashkent celebrated International Women's Day by proclaiming hudjum, an attack on old o6ychea and traditions aimed at 6oee globally changing everyday life and relations between the sexes. One of the main goals of hudjum was to eliminate the burqa (head-to-toe veil) worn by urban Muslim women and girls in the presence of unrelated men.

The programme should have been completed in less than six months in time for the commemoration of the tenth anniversary of the 6olshevik revolution. Instead, it took three decades and another major upheaval, the Second a world war - in order to make a widespread "pazo6lachnoe" and The "ocw6o6ooedding" of Central Asian women has taken root.Meanwhile, the women found themselves under pressure from two sides: the Soviet authorities on the one hand, and the persecution, especially of their families and neighbours, on the other. At home, they risked severe abuse, divorce, and even murder for exercising their new rights. On the outside they could be treated as prostitutes. In the early years of the hujum, thousands of Central Asian women were attacked and raped, and hundreds were y6xualized, often by their male relatives. It is not surprising that many women who initially burnt their burqas were later unobtrusively exposed, either by replacing the veil with a shawl or by lowering the veil and revealing it, depending on where they were. Social propaganda campaigns to end the veil, kalima and other repressive practices in the region continued until the 1970s.In 1930, the Soviet government began a campaign for universal primary education. Within ten years, enrollment in schools in the Soviet Union tripled, and approximately one third of the enrollment was in non-Russian pec6y6 lics. This upsurge has put forth a major initiative to recruit and o6ypecialize teachers locally. The number of teachers in the Soviet Union as a whole grew from less than 400,000 in 1928 to over one million by 1939. Women played a crucial role in filling this niche. Throughout the CCCP, not least in Central Asia, teaching would become a predominantly female profession.Under Soviet rule, Central Asian artistic traditions were fused with the techniques of Soviet art - "national in form, socialist in content". In addition to encouraging the preservation of traditional arts and crafts, such as copper engraving and carpet weaving, the Soviets - in what many today consider an act of cultural imperialism - imported and encouraged Western-style art forms. Plays by Russian and Western authors were staged. Conservatories were opened and classical musicians began to be trained. Some Central Asian women became artists, actresses, musicians, and dancers. The daughters of mothers who had recently been given a chance to perform traditional dances, previously reserved for men, while others mastered such Western styles This bold break with tradition was made possible in large part by the vigorous and comprehensive education of young people in the Soviet school system. Throughout the CCCP, young people attended kindergartens and schools that looked alike, with the same curriculum and the same textbooks in Russian and national languages. All primary school pupils were enrolled in the youth troop "Oktia6picty" (named after the Great October), third-graders became pioneers, and high-schoolers joined the Komsomol.In the summer, they were additionally socialized in summer camps built on the o6paize of the famous Artek camp on the Black Sea. These mostly co-educational settings were a major departure from the gender segregation that had prevailed in the region. They inculcated in the young generation a new, 6olie egalitarian system of values, art forms such as 6ault and theatre.

For a long time, the "women's issue" in the Soviet Union was considered a priority, as the Soviet government pursued the goal of attracting women to the highest levels of state power. After the end of the Great Patriotic War, reforms concerning the economy, politics and social protection of the population were issued for decades. In the 1950s and 1960s, women in the sphere of material production were included; a large number of women were also registered in both urban and rural areas, and agriculture was enriched with female personnel. All this has been accompanied by social assistance from the state and a set of measures designed to support women in their careers. Thus, Article 35 of the 1977 Constitution of the Soviet Union not only contained equal opportunities of access to o6pao nition and professional training, employment, remuneration and promotion, social policy and cultural activities, but also special measures of labour protection. and health. for women: conditions that allow women to combine work with motherhood; legal protection, financial and moral support for mothers and children, including paid leave and other benefits for pregnant women and mothers, as well as a gradual reduction of working hours for women with young children [9].

The above-mentioned judgment that the "women's" issue in the Soviet Union was over is also explained by the consequences of the Great Patriotic War [10].

According to statistics, over 700,000 women in the Soviet Union willingly fought at the front together with men. In the 1940s, 70 per cent of medical personnel in the CCCP were women, and almost 90 per cent of liaison officers in military units were also women. During the war years, a great deal of assistance was provided by women who participated in partisan unit formations[11].

One of the negative consequences of the Great Patriotic War was a decrease in the birth rate in the countries, huge demographic losses, material losses and a decline in industrial production. Because many men who fought at the front were killed or disappeared, it was extremely difficult to restore the birth rate. The decline in the male population in the CCCP has created difficult conditions for family survival. Women, in order to provide for their children, have therefore been forced to work in traditionally "non-feminine" occupations. B In the decades following the end of World War II, the CCCP verdict was that women had equal rights with men because they had the same conditions to work, earn money and develop themselves. At that time, the view was developed that helping women was an essential part of building the socialist state. Historians described it in no other way than

"the destruction of the capitalist way of life", "the development of the 6e classes" and "the peo6pa6pa6onization and ocuvrati6on of the common people" [12]. Thus, the certainty of the judgement that the "female" question in the Soviet Union was finally resolved in view of the fact that in the 40-60s women were an important link in the process of industrial production, and there were no cases of gender discrimination with regard to their occupation. Mechanisms for solving this problem must reach a combination of the three statuses for Soviet women: worker, mother, and citizen. The equality of rights with men in those times was justified by their financial independence, the woman herself established herself as an individual at the expense of "creative labour". Researchers have noted that during this period of time women raised their level and authority in the family at the expense of labor. A sense of responsibility was being formed, of fulfilling one's duty to the fatherland.Under socialism, any woman had the right to pursue her own occupation as a worker in production and other labour activities. In the 1980s, about 140 million women lived in the Soviet Union, almost 55 per cent of the total population. It is important to mention that agriculture was one of the most female-employed sectors, employing about 45 per cent of women in all Soviet occupations. This percentage was relevant for a whole decade and was not characterized by any potential decrease [13]. Soviet sociologists argued that women in the Soviet Union were not only motivated to work for wages, but also motivated by a moral duty to the country and a desire to participate in public and political life.
The scientific and technological revolution has had a significant impact on the conditions and development of women's work. The mechanization and automation of production processes and the increase in the quality of scientific organization of work have led to the expansion of the boundaries of women's work. It has thus had the effect of increasing the percentage of women's participation in engineering, electrical engineering, manufacturing and other high-tech industries. As a result, one out of every four women in the Soviet Union was an engineer, and three out of every five doctors were women. The female population was more interested in scientific and educational activities. Positive results were also seen in the field of education: 75% of teachers in educational institutions were women, and about 92% of them were in charge of primary classes. However, despite the above-mentioned results, only 36% of school principals in the CCCP were women. Similar cases were common in other fields, and the 26th Congress of the Communist Party of the Soviet Union subsequently concluded that the society was not yet ready for women to occupy managerial positions, and that there was insufficient opportunity to promote women to higher positions. The congress was instructed to implement measures

to remedy this situation [14].
In Central Asia, before the advent of Soviet rule, women's lives were regulated either by Muslim law (Shariah) or by local national law. In general, the way the women of the region developed was rather uniform: at an early age they were married, where they were regarded as full co-contractors of their husbands and families.In the Soviet Union, the number of women who worked in manufacturing increased dramatically in Central Asia as well as in other countries of the Union. For example, between 1930 and 1980, the proportion of women workers in Central Asian countries increased as follows: in the Kyrgyz CCP from 10% to 49%, i.e. almost 5-fold, in the Yz6ekkoi CCP from 20% to 45%, i.e. about 2.5-fold, in the Turkmen CCP from 20% to 40%, i.e. 2-fold, in the Tajik CCP from 8% to 40%, i.e. 5-fold, and in the Kazakh CCP from 15% to 50%, i.e. 4-fold [15].
The government of the Soviet Union not only made provisions for Soviet women to work in industrial production, but also in the public and political sphere. Under the CCCP Constitution, women on an equal basis with men had the right to occupy high-level posts connected with state management. Inclusion of women in public affairs was an unusual innovation for men, particularly in Central Asia. Even in pre-Soviet Russia, women were not allowed to participate in the executive and judicial branches of government. With these innovations, more women in Central Asian countries began to occupy positions in such bodies as courts, prosecutors, responsible positions in trade union and Komsomol bodies, in parties, and so on. The highest results were in the Tajik, Kazakh and Kyrgyz CCPs. In Tajikistan in 1979, 21% of women held party posts and 43% held people's sessions. In Kyrgyzstan, 27 per cent of party members were women, and women were about 51 per cent of lay judges, respectively. In Kazakhstan, these figures were as follows: 52,3 % и 59 %. In general, half of the members of people's assessors in all Soviet republics were women.
The Soviet Union also considered important factors such as motherhood, so it created conditions for mothers who were also mothers. For example, in the CCCP legislation there were clauses stating that there were some factors which depended on the psychological and physical components of men and women. Therefore, there were some professions that were not suitable and prohibited for women in the CCCP. These included working in the chemical, printing and leather goods industries, as the conditions of these professions were very harmful to women's bodies. It was also forbidden to allow women to work underground, in mines, etc. Thus, the CCCP did not strive for a6co-litual equality between men and women, based on the oco6eonnectivity of the

organism of the o6oic sexes.During the existence of the CCCP, some 30,000 organizations and advisory bodies responsible for women's work were established. Their function was to provide women with psychological and moral assistance. Women and children received medical care from the state throughout.
Moreover, the Soviet Union also predetermined conditions for women in childbirth. For example, a woman on maternity leave received financial assistance from the government in the form of fully paid leave, as well as payment for the care of her child, which could last for a maximum of six months. Pa6odtors were instructed not to dismiss pregnant women or women who were bringing up children under two years of age. The birth of children not only brought financial assistance to the woman, it preserved her work record.
At various congresses of the Central Committee of the Communist Party of the Soviet Union, the question was often raised of increasing the competitiveness of women in the workplace, of improving all conditions for them, taking into account, above all, the physiological and psychological characteristics of the female body. Furthermore, it is important to improve the mechanization of labor-intensive manual labor, fully removing women from heavy and other unhealthy tasks, and then completely eliminating women's work from night shifts [16].
The conditions and processes of women workers were monitored by various agencies, special trade unions, the Ministry of Labour and other authorized organizations specializing in women's work.Studies show that the higher the level of competence of the spouses, the more the husband participates in household chores. In the Soviet Union, there were more young families than families with older spouses. All this confirms a trend in family relations, which is very favourable for women. But nationally, women still spend more time and resources on their families than men do. That is why the efforts of the state to transfer some functions of family services to state institutions of everyday o6culty are aimed at reducing women's domestic work, so that Soviet women could do it in the first place. It was very important for the Soviet state to build all conditions for women to combine opportunities to raise children and to participate more in o6present life, o6paecoise and get the rest they deserve.At the same time, it should be noted that in the CCCP the state support of the family was still far from its proper level, there was a need for further development and a significant qualitative improvement in the work of state and private enterprises and trading companies.At the same time, it should be noted that some unintended consequences are associated with a tendency to grant only women, and not one of their spouses, the right to a short holiday and to receive a

certificate when their children are ill. The woman was able to leave the job until her child was two years old, take care of household chores and receive social public assistance and two years of leave and other benefits for not having a full child. The challenge was to arrive at justice in this crucially important issue for true equality between men and women, and it was necessary to eradicate the lack of equality, as it had a negative impact on women's productive and political activities and on the development of equal relationships within the family.The tendency of increasing women's participation in political activity in the XX century was peculiar not only to the Soviet Union, but also to the West European countries, which was a factor of moral interest of women in social problems, of accessibility of the given issues for them, of their self-development and self-consciousness growth. B. I. I. Lenin asserted that among ordinary women-pa6o women there were also many personalities who had the potential for managerial functions and a copo6ocy for responsibility and organisation.

Marxist-Leninist ideology on "women's issues" and Soviet legal guarantees created a gender paradox that, despite high achievements in o6licity of education and employment, reflected a 6oee low level of women's participation in political institutions. This is a legacy of the Soviet system that contemporary Central Asia needs to address in responding to the challenges of political empowerment in the transition to democracy. In response to international interference and the globalization of gender mainstreaming, the governments of most post-Soviet countries have taken measures to empower women politically. Notably, however, the changes brought about by the transition to democracy in Central Asia have had a negative impact on gender equality. Given the cultural and structural changes in attitudes and norms of everyday life, as well as the male-dominated political culture, the success of these policies depends in part on the ability of Russian women to meet the challenges through new forms of 6op6 advocacy for shaping the political process in their favor.

1.3. The situation of modern women in the Central Asian region

Any development and progress in the political and legal life of the society is unattainable if it does not involve changes that affect specific categories of the population, minorities and other groups of society. Underestimating the transformation of gender policy is a direct path to the destruction of democratic institutions and the ineffectiveness of civil society, it causes mistrust on the part of the public, and it also provokes confusion among the population. That is also true for women. Clearly, women's rights to participate in political power were established relatively recently around the world, and different developments are still taking place. Despite the fact that in the world there is a process of active inclusion of women in the political and legal way of life, the gender factor that generates stereotypes and negative consequences is still relevant. According to global statistics, there are trends related to women in politics. For example, women's participation in electoral power is many times less than men's. The same can be said about women's interest in political processes. This is due to the fact that the main organs of power have historically been established and formed by men. Even the census in the 18th century included only men.

When women take part in presidential races, the following happens: their campaigns are not as intensively promoted as men's programs, and the population, in principle, does not have the same trust in their candidacies. In world practice, there were and still are many cases in which women, in order to preserve their authority and political status, have to adapt to the traditionally "male" political environment. These cases are relevant even in Western countries, which are globally recognized as an advanced and tolerant society, where the promotion of women's rights and opportunities among the masses is considered a pacpopular and o6efficient practice, especially in Eastern countries, where the leadership positions are traditionally occupied by men.

Despite this, statistics show that women are more active in campaign activities. More women than men prefer to collect candidate support signatures, support the work of the Electoral Commission, and contribute financially to campaign funds.

Women's participation in political processes is considered a non-traditional political experience. Considering this judgment, women who hold high office always tend to raise very specific but unimaginable issues that require immediate solutions: domestic violence, gender equality, trafficking of women, rape and so on.It is common in Central Asia, as in the rest of the world, for women to be underrepresented in political office. The constitution of each of the five Central Asian member states states states that regardless of gender, religion, ethnicity or nationality, every citizen has a guarantee of eligibility for political participation and positions. According to the latest figures from the International Parliamentary Union, only 20 per cent of women around the world participate in political development, hold parliamentary seats and high-ranking positions in relevant bodies. Achieving equity in the successful implementation of gender policies in the Central Asian region (namely, the intensive inclusion of women in politics).

According to "Women as Agents of Change in Central Asia" by P. Turaeva, who is an anthropologist with a PhD in gender issues in the Central Asian region, significant changes are currently taking place in the region: women have become more open to their contribution to political and legal institutions and have become more active and proactive. For example, such fields as local economics and self-government are nowadays actively promoted by women, and women's entrepreneurship is spreading. It should be taken into account that Islam plays a dominant role in shaping the consciousness of individuals in almost every Central Asian country, and women's rights to self-determination and career development in Islam are quite limited. It is believed that after the long propaganda of atheism in the Soviet Union, people began to return to their traditional roots, namely Muslim values. Sometimes people perceive this negatively, i.e. as promoting discrimination against women. However, even religious teachings can be used for business purposes.

Despite this, women's empowerment is seen every year and the prognosis is very optimistic: change is gradual but effective.

Turaeva writes that in academic coo6cientists it is commonly believed that after the collapse of the Soviet Union very negative processes took place in post-Soviet countries: shortage of jobs, decline in industrial production, reduction of enterprises, deterioration of infrastructure and so on. For these reasons, this difficult period is referred to in research circles as 'de-modernisation'. Inhabitants of Central Asia remember this period as the "Stone Age" of poverty, hunger and criminality. The above-mentioned trends of the new post-Soviet era have brought not only difficulties and predicaments, but also metamorphosis of society and new opportunities. The author argues that with the building of democratic institutions in the newly independent states, the consciousness of the population grew, so that women started to think more about their rights and prospects in new spheres, which had not been possible for them for a long time. Turaeva observes that throughout Central Asia there is little scholarly research on women's leadership, their aspirations and plans, the development of female entrepreneurship, and other areas. The anthropologist believes that nowadays women are viewed only through a black prism of judgments: whether a woman is concerned with everyday life or career opportunities; a woman is viewed in a context where she is positioned as either strong or weak.

Despite women's initiative, traditions, stereotypes and prejudices still play a strong role in Central Asian countries and are a deterrent to a great extent. After the collapse of the CCCP and independence, the states began to return to that legacy and previously established values. Traditionally, Central Asia is a region with an inherently patriarchal way of life, where women have to adapt, navigating between accepted frameworks and patterns and new trends, in order to secure a secure future for themselves and their children. As a consequence, nuances gradually emerged in the formation of women's institutions. For example, the emergence of Women's Councils in Soviet times heralded the beginning of an increased role of women in political and legal bodies and was created by the Soviet authorities to address gender inequalities. [17] In this way, the government wanted to rid the populace of the entrenched and irrelevant judgement that women's main task was procreation and the 6yctoffectiveness of 6their life. Fighting against patriarchy, the Women's Councils encouraged women's activities outside the home, their pursuit of higher education and so on. However, these bodies fulfilled a different purpose: controlling women's behaviour, particularly in institutions of higher learning. Thus, the Women's Councils, which were originally created to promote equality between men and women, have set up surveillance on female students and have imposed fines for some breaches of university rules of conduct. Unfortunately, such cases continue to this day.

In her work, Turaeva notes that the concept of a "modern woman" is totally different in the Central Asian states. As for the Soviet period, a "modern woman" was considered to be one who ocurred from patriarchal prejudices and did not consider it her duty to devote all her resources and time to creating a family and the like. The author believes that, at this point in time, such The fact that in some countries of Central Asia life is now criticized and generally perceived rather negatively means that society is not ready to see women as fully formed individuals in all respects. At the same time, Turaeva writes that the role of higher education in the lives of young girls is assessed quite positively, and virtually every girl, after receiving her secondary school diploma, passes her exams to enter a higher education institution. The enrollment of girls in higher education is in countries such as Kazakhstan and Kyrgyzstan, according to the article"The gender gap in tertiary education in Central Asia, by E. Ca6zalieva, since the 1960s, the number of female students in universities has exceeded the number of undergraduates. Of all the countries in the region, Tajikistan has the least number of girls enrolled in higher education, with only 40% of schoolgirls going to university. Nevertheless, statistics show that after marriage, many women, even with a higher education degree, are forced to leave the field. Thus, in Central Asia, higher education for women is more formal, an indicator of prestige rather than a guarantee of employment.

Turaeva also writes that it is paradoxical that women want to preserve and develop traditions that are specific to young daughters-in-law. In Central Asian states, the daughter-in-law, also known as "kelin", is the face of the family; she must have a good reputation and conduct herself in a proper way, from upbringing to socially acceptable clothing. At the same time, there is an opinion that the "daughter-in-law" has the lowest rung in the family hierarchy. This is due to the fact that a married girl has to do maximum household chores, to be submissive to her husband and his parents, always friendly and ready to fulfil her domestic duties. As a rule, it is very difficult for a woman in the "kelin" status to pursue her career growth, as it rarely involves a new family, and maneuvering between family and work is almost an impossible task for many women.

In general, Turaeva notes that women are innovating in spite of traditions and other constraints, such as religiosity, the usual patriarchal way of life and so on. In the region, there is a fairly active process of glo6a lization, political changes, l6epa lization, growth of organization, which has a positive impact on gender issues. Conservatism, which coexists with new trends, is mainly a negative factor. The researcher writes that the neo6xoocity of research pa6 otic on women's goals and perspectives exists in all Central Asian states and a focus on these issues will affect progressive change.One of the current ppo6lems of this region in relation to women is violence and the perception of violence in o6ececc tion as a reasonable and purposeful neo6xo odimocity. Violence is one of the most important negative trends that must be completely and universally eradicated. It not only prevents women from living and developing as persons, from thinking fully about their families, but also from engaging fully in other activities and building a civilized, developed society. Taking violence in the community as a given has long term detrimental effects not only on the community, but also on countries and international society as a whole. About 30% of countries still, in the 21st century, have no protection for victims of domestic violence in their legislation. Risk factors for exposure to violence are: 1) low quality or no quality at all; 2) tolerance of all types of violence in society; 3) 6espa6o ticism; 4) economic and social crises; 5) acceptance of gender inequality and lack of eradication as such and other factors. Every state loses about 4% of its GDP because of 6youthviolence. This figure is twice as much as most states spend on areas such as education, culture and sports. This is one of the problems that keep Central Asia from reaching the final outcome of gender equality. In the opinion of 6oolscee ment of sociologists, in the region of Central Asia there is still an atmosphere of houseekeeping in families, where women endure humiliation both physically and morally, keeping it a secret, under fear of shame and lack of guarantee of real help.

Historically, women, under the influence of Oriental mentality, have long carried the cross of an apathetic member of the social hierarchy whose function has included, for the most part, the household.

According to the OOH gender equality and women's empowerment framework called "OOH-Women", one in three women in Central Asia has experienced various types of discrimination. This includes crimes such as sexual rape, rape, psychological violence and so on. Moreover, the above-mentioned and inherent traditionalism and conservatism in Central Asia tend to distort many religious and national attitudes, justifying violence and preventing women from taking responsibility for their own decisions and lives. Also, in the region there is a lack of awareness among women about their rights and opportunities. According to Human Rights Watch, a non-governmental organization, women in Central Asia who have been subjected to sexual violence receive insufficient assistance and support, and subsequently receive justice from and further protection from their abuser.

For example, in Kazakhstan, the Law on the Prevention of Domestic Violence was adopted in December 2009, but the government has a long way to goA painstaking effort has been made to eliminate legal shortcomings that prevent victims from receiving adequate protection. This law has defined various types of violence and established penalties for committing these crimes. Victims are entitled to receive social services for protection and support. However, the Penal Code of the PK does not even mention the criminal responsibility for domestic violence. In almost 10 years Kazakhstan should have been very successful in this area, but in fact the statistics show sad results: about 17% of women aged 17 to 77 in Kazakhstan have experienced physical or sexual violence.

Every year, about 400 women are victims of domestic violence. About 15,000 calls a year are received by the "Union of Crisis Centers" asking for help, 90% of them from women. In fact, domestic violence is still perceived by the public as a "family matter" that should not be interfered with. Also, statistics do not show all the real numbers, due to the silencing of the crime. The main problems are that the police do not adequately inform the victims about the social protection mechanisms. Human Right Watch has also stated that the police often influence women to withdraw the case from court and make peace with the perpetrator. In addition, there is a contradictory trend in Kazakhstan of placing the blame on the victim while excusing the aggressor. In 2020, Channel 31 launched a talk show called "Kel, tatulasaiy! ("Let's make peace!"). The purpose of this program is for the victim to forgive and reconcile with the person who caused her harm, and there are frequent instances of the victim being ashamed of having brought a criminal case against the perpetrator. In spite of the public outcry and indignation, the talk show is still aired on Kazakh television, instilling in the minds of people the idea of humility with the state of affairs and the need to recognize victims of violence.

Another o6schewed resonance in Kazakhstan was caused by an innovation adopted in July 2017. Former PK President Nursultan A. Hazap6aeu signed a new law on the decriminalization of
"assault" and "intentional infliction of minor injury". These articles are mainly used in cases of domestic violence. As a result of this law, it has become even more difficult for victims to initiate criminal proceedings against their aggressor. In January 2021, the current President of Kazakhstan, Kassym-Zhomart Tokayev, spoke out about domestic violence. The President said that the protection of women is one of the top priorities in the country, so the new draft law "On counteracting domestic violence", which had previously been misunderstood by Kazakhstanis andmet negatively, will be reconsidered by Members of Parliament[18].

In the case of Tajikistan, they had already in the Soviet Union begun to take a careful approach to this issue. For example, it was one of the first states in Central Asia to legislate the promotion of women to power. In Tajikistan, the inclusion of women in the development of politics, the economy, business, and local self-government has long been a matter of national concern. In 2001, the government adopted a long-term programme to improve gender equality in the country.Moreover, the program included many points and tasks to increase the role and status of women in society. In particular, a decisive measure was a 30 per cent quota for participation in the legislative, executive and judicial branches of government. A similar quota was granted by President Emomali Rahmon of Tajikistan for girls living in rural and remote mountainous areas to enrol in higher education. Despite progressive innovations and the state's commitment to a new course of development, the influence of Shariah law and national traditions have proven too great to start a new stage for women. At the moment, about 13 per cent of women in Tajikistan hold parliamentary seats with the status of deputy. Unfortunately, the aforementioned quota of 30% is unknown for many regions of the country, and women remain unaware of their opportunities. Sociological analyses have shown that a clear deterrent to progress is the radical patriarchal structure. Also, Tajikistan is a state where the majority of the population is Muslim. The percentage of women who are subject to moral and physical violence and all kinds of discrimination and disadvantages is 70%. To cope with this problem, a sufficient number of non-governmental organizations operate in Tajikistan with the aim of protecting women from arbitrary violence and cruelty. Their main help consists in legal and spiritual education of the female population. There are also crisis centres for abused women throughout the country. According to the staff of these centres, the number of calls does not decrease every year; on the contrary, it only increases. The number of women in need of psychological assistance after an incident is also either increasing or decreasing every year.

In Turkmenistan, the situation of women is much worse. Turkmen law formally stipulates equality between men and women and also guarantees the full protection and inviolability of women from violence. Turkmenistan was one of the countries that acceded to and signed the Convention on the Elimination of All Forms of Violence against Women in 1996. In fact,the real picture does not coincide with the measures taken by the authorities [19].

One of the reasons why this topic is not covered as much as it should be is because of the closed nature of the state. Official statistics about women victims of domestic violence are hidden. Despite this, according to local residents, journalists and sociologists, domestic violence is an uncommon practice. Most Turkmen women face such things as abuse and coercion. In many families, women provide for their own families, for example, by sewing clothes and carpets at home.

After the collapse of the Soviet Union, a crisis started in Turkmenistan, the consequences of which were the closure of many enterprises, organizations and institutions, which led to mass emigration. At that time, emigration took place: men left for other countries (mainly Russia, Turkey and the United Arab Emirates) to look for work. As a consequence, the male population in Turkmenistan, who were the breadwinners of their families, decreased in the 1990s. The Turkmen women had to carry the burden of o6e 6people to support their families. There have been frequent cases of women having two or three spouses and the birth of outside children. To this day, women have occupied the niche of traditionally "male" labour, selling various kinds of goods in markets.

The government does not show the real picture, not only because of patriarchy, but also because of a lack of real up-to-date data. Women in Turkmenistan are intimidated by their husbands to keep quiet about things that have happened. "Don't take the cop out of it" is one of the principles of the Turkmen mentality.

Going to the police, too

- unspoken ta6y and an embarrassment to women. Even though non-governmental organizations exist, in practice they are ineffective for the reasons mentioned above.

In January 2008, the United Nations task force on periodic review stated that Turkmenistan was not addressing gender inequalities and was not fully aware of the consequences, and its legislation did not establish preventive measures against violence and social assistance to protect victims [20].

Regarding the activation of women in Turkmenistan, mainly women from wealthy families have created conditions for women's entrepreneurship. The main institutions where Turkmen women participate are trade unions of entrepreneurs and industrialists, newspaper and magazine editorial offices, o6paozevatelnye organizations and so on. There is also the Centre for Business Women of Turkmenistan, which aims to improve the qualifications of women working, inter alia, in the agricultural industry. It supports women farmers by providing them with all necessary inputs to promote theirproducts of activity. Turkmenistan also plans to create a similar organization to support women entrepreneurs. In Turkmenistan, it is the service sector that is most developed among women who are engaged in trade. Construction, greenhouses and so forth have also become more widespread among women in recent years. The government supports these projects through credit programs.

In Yukitania, according to the latest data from the Institute for War and Peace Reporting, there has been an increase in the incidence of female forced marriages among the female population. That's not due to financial problems at all, former Deputy Prime Minister Dil Gulyamova told Central Asia, but mainly to the violence of husbands and domestic partners, she said. Official statistics, similar to those for Turkmenistan, are not available. All that is known is that since the start of the pandemic the level of violence has greatly increased. According to information from the Ministry of Internal Affairs, between January and December 2020, some 9,000 protection orders were issued in the country to provide protection to female victims. Of these, 60% were physically and 40% were psychologically discriminated against.

The United States, until recently, was a closed state with a policy of isolation in the international arena, hence, the latest trends were nothing to them. Since 2018, Yoz6ektan has invested more resources and attention in eradicating gender inequalities and promoting the role of women in Yoz6ektan society. Parliament, along with the government, has been introducing and implementing reforms that include economic and social opportunities for women's career paths. An important pillar of these innovations are the two laws adopted in 2019:

"On guarantees of equal rights and opportunities for women and men" and "On protection of women against harassment and violence". The latter law is innovative in that it provides a legal basis for the protection of women victims of domestic violence, as well as further assistance in criminal proceedings and, eventually, in the court itself. Violence has always been a serious criminal offence in Yugoslavia, but in the past there was no documentation of such cases by the Ministry of Internal Affairs and no social mechanisms to provide assistance and support to women victims. At the moment, in Uzbekistan, if an abusive crime with physical harm to a woman occurs, she will subsequently be provided with medical, psychological and economic assistance, which will also be paid for. Also, at the beginning of 2020, protection orders for women victims of gender-based violence were introduced. Such measures will help to control their physical 6effectiveness. These reforms were fullysponsored by the World Bank, it is part of its plans for new social and economic initiatives.A social and legal aid organization for women and children, called "Mehjon", operates in Bukhara. Its activities are carried out in accordance with the local authorities and the government. Since the beginning of the quarantine in the country, the cases of violence have increased manifold. According to the head of the center, there were about 300 cases during the first months. During the months of self-imposed isolation, from March to May 2020, about 600 protection orders for female victims were issued. It is worth mentioning that in the second half of last year 2019 there were no cases of these warrants, i.e. there were no cases of domestic violence on such a large scale [21].

Kyrgyzstan, which in recent decades has been characterized by unpredictable shifts in power, social upheaval and economic crisis, has also kept pace with its neighbors in the region and is gradually moving toward progressive laws that protect and support women. As in other countries, the number of crimes against women increased in Kyrgyzstan in 2020, but Kyrgyz human rights activists argue that the figures are unrealistic, as the mentality will not allow women to go to the police with their ploys. In 2019, the former Prime Minister of Kyrgyzstan, Mukhamedkali A6ylgaziev, delivered a message to the people, in which he said that measures to toughen penalties for aggressors are unnecessary and said that the government is yo6y6ing activities to protect women's rights. As this ppo6eema has long 6ecpoekedo o6ececoilo, the government has also poo6ed to take action and, in 2019, has officially released statistics on the victims. In January 2019, the deputy head of the Ministry of Internal Affairs spoke on a similar topic, in which he criticized the work of law enforcement agencies. According to Kyrgyzstan's latest statistics on cases of 6youth violence, they have been recorded as the most frequent over the last 10 years. Authorities have registered about 10,000 reports to the police, most of which have been from women. A striking fact is that about 89% of the cases were subsequently frozen and withdrawn. As in Kazakhstan, there is also a practice in Kyrgyzstan of influencing women to choose to reconcile with their abuser rather than to continue to oppose their abusers' actions. According to the Ministry of Internal Affairs, most criminal cases are not fully clarified because the victims, under pressure, also decide to withdraw under the pretext of conflict resolution. It also happens that the rapist writes a counter-report against the victim.

If we look at Kyrgyzstan's statistics for the last 10 years, 97% of the perpetrators of serious crimes are men. Most of them are wife abusers. Mostly men commit abusive acts,under the influence of alcohol or drugs. Only 5% of rapists are brought to criminal responsibility. Bespa6oticism is also noted by the Ministry of Internal Affairs of Kyrgyzstan as one of the triggering characteristics of criminals. In 2018, a sociological study was conducted in Kyrgyzstan, which showed that many women suffer from victimized behavior. This term o6o ncludes the victim's propensity to get into situations where she becomes a victim. Based on the results, out of 175 women in Kyrgyzstan, only three of them o6paaatec to law enforcement agencies.

It is worth noting that most of the aggressors get away with mostly administrative penalties. According to the above-mentioned fact that most rapists are guilty, the administrative fine should eventually be paid by the victim, not by the perpetrator. In the end, this fine goes to the state treasury, not as compensation for the victim. According to Tolkun Tulekova, executive director of the Association of Crisis Centers, administrative liability is harmful because violence is not eliminated, but rather it leads to the original point, i.e. it becomes a vicious circle.

2. POLITICAL LEADERSHIP IN CENTRAL ASIAN COUNTRIES

2.1. The current legal and regulatory framework for women's equality

One of the first international organizations to be committed to protecting women's rights and rights was the United Nations. Numerous conventions and instruments related to gender equality policies have been signed and ratified by the OOH since the 20th century. The very first were the lines in the organisation's most important document, the Bceo6o te Declaration of Human Rights, which states that a6coo lute all human beings are born, 6young and have equality in their rights and entitlements. About 100 OOH documents deal specifically with equality between the sexes. These include: "The Convention concerning Trafficking and the Exploitation of the Prostitution of Others (1949); MOT (International Labour Organization) Convention concerning Equal Remuneration for Men and Women Workers for Work of Equal Value (1951); Convention on the Political Rights of Women (1952); Convention on the Nationality of Married Women (1957); Convention against Discrimination in Respect of Employment (1960); The Convention on Consent to Marriage, Minimum Age for Marriage and Registration of Married Persons (1962); the International Covenants on Economic, Social and Cultural Rights and on Civil and Political Rights (1966); the Optional Protocols to the International Covenant on Civil and Political Rights; the Convention on the Elimination of all forms of Discrimination against Women (1979); MOT Convention 156 on Equal Opportunities and Equal Treatment for Men and Women Workers: Workers with Family O6auntsy" (1983); Convention against Torture and Other Cruel, Inhuman or Degrading Treatment or Punishment (1987); Convention on the Protection of the Rights of All Migrant Workers and Members of Their Families (1990); Optional Protocol to the Convention on the Elimination of All Forms of Discrimination against Women (1999) and other instruments".The United Nations has defined gender mainstreaming as: "...the process of assessing any planned action based on its impact on women and men, including legislation, policies and plans in different o6lacations and at different levels. The strategy is based on the fact that women's and men's interests and experiences have become indispensable standards for conceptualizing, implementing, monitoring and evaluating all political and policy work. Economic and social o6lacities enable men and women to benefit equally, and inequalities will never 6yearn".It is important to mention that all of the five Central Asian countries have ratified all the documents of this OOH policy, as well as support and strive to create all conditions for equality and justice.

Each of the Central Asian republics has articles in their constitutions stating equality before the law and the courts, state guarantees of equal human and civil rights and freedoms, including regardless of gender. In Kazakhstan, the President signed the "State Guarantee of Equal Rights and Equal Opportunities for Men and Women" in early 2009.

Kazakhs and Kazakh women now have equal rights to engage in o6social activities, but women still receive certain work benefits after sentencing and are not obliged to serve in the army. Bo all other things in Kazakhstan pa6oate the principle of gender equality.

In terms of promoting gender policies, Kazakhstan was the third country in Central Asia to sign laws on gender equality. Tajikistan and Kyrgyzstan were pioneers in this area.

In Tajikistan there is a similar law "State guarantee of equal rights of men and women and equal opportunities for their realization", the task of regulating relations on equal rights of men and women, guaranteed by the Constitution of the Republic of Tajikistan. Tajikistan from the very o6pee tion of independence has sought to prevent gender discrimination in all spheres of social, political, cultural and other spheres, o6ecpee ning equal opportunities for people of different sexes.[24]

As for Kyrgyzstan, their similar law is the "Law on Basic State Guarantees on Guarantees of Gender Equality", which regulates relations with the aim of providing equal rights and opportunities for people of different sexes in social, political, social and political, economic, cultural and other spheres of human life. This law protects men and women from discrimination on the basis of sex. It seeks to establish progressive, democratic relations between men and women based on national traditions. It provides a State guarantee of equality. Kazakh legislation in principle provides for the same.

At the same time, if we compare all three laws regulating legal relations of the three countries, the most progressive is the Kyrgyz law. This law has already been amended and corrected, that is, it has been edited for the second time. This gives grounds to believe that the Kyrgyz law is not only declarative. For example, the Kyrgyz version of the document deals with the issue of quotas in the sphere of public administration.

"State and local government quota system (not 6oolee 70% of employees, including decisions - creation level) is established by legal and regulatory acts of the Kyrgyz Pecuny6lyk"(Art. 10). Also mentioned are the provisions on the field of labour relations."Pa6oeditors are prohibited from advertising (advertisements) in which a vacancy offers pa6oety only to women or only to men. With the exception of certain tasks that can be carried out exclusively by a person of a particular gender offering different conditions and requesting information about people seeking pa6oety, personal life and birth plans" (Article 18).In the countries of Central Asia, positive changes and a desire for liberalization in all spheres of life came in the 1990s. At that time the political formation of civil society, its development and democratization began. With the emergence of pluralism and people's activism came new democratic institutions that promoted gender equality.In recent years, Kazakhstan has become one of the leading countries in the region, where women's rights and the conditions of their life and self-realization have been improved.Kazakhstan has ratified all the major universal conventions on gender equality.These include the United Nations Convention on the Elimination of All Forms of Discrimination against Women, the civil status of married women and women's political rights, civil and political rights and economic, social and cultural rights.Equality of rights and rights of men and women in Peculiar Republic of Kazakhstan is a constitution, i.e. one of the basic human rights provided for in the Constitution of Peculiar Republic of Kazakhstan.Thus, in accordance with the Constitution of the Republic of Kazakhstan, Kazakhstan recognizes and guarantees human rights and freedoms (Article 12(1) of the Constitution of the Republic of Kazakhstan). They extend to every person from birth and are deemed to be a6coollateral and indivisible (Article 12(2) of the Constitution of the Republic of Kazakhstan). This applies to the rights and entitlements of men and women.The Constitution of Kazakhstan proclaims the equality of all before the law and the courts (article 14, paragraph 1, of the Constitution) and prohibits any form of discrimination on the basis of sex (article 14, paragraph 2, of the Constitution).

The Constitution also contains provisions guaranteeing the implementation of human and citizens' rights (art. 13, para. 2; art. 15, para. 1; art. 16; art. 17; art. 18, para. 1, para. 2; art. 19. ; pp. 1, pp. 2;art. 21; art. 22, p. 1; pp. 24; pp. 26; art. 28, p. 1; pp. 1, 2 Art.f the Constitution of the Republic of Kazakhstan; Article 30, para. 2; Article 33, paragraph 1; Article 4, paragraph 2 of the Constitution of Pecpy6lika Kazakhstan) and its protection (Article 1, paragraph 2 of the Constitution In this regard, it can be said that the current Constitution of Pecpy6lika Kazakhstan B Kazakhstan has the principle of equality of men and women in particular, in accordance with Article 1. Article 24 of the Constitution of the Republic of Kazakhstan stipulates that "everyone shall have the right to a cov6 ood work and to a cov6 ood occupation and profession", which is based on the provisions of Article 3 of this Article. Article 24 of the Constitution of the Republic of Kazakhstan states that "everyone has the right to remuneration for work without discrimination of any kind" in accordance with Article 4 of this Article. 33 'Citizens of Pecuny6lyk have equal rights to public services'.Thus, the Constitution grants men and women only na6op rights that require not only recognition and compliance, but also appropriate measures taken by the State to realize those rights.It should be noted that many constitutional provisions have been enacted into law, in particular, the Law on Labor of the Republic of Kazakhstan, the Constitution of the Republic of Kazakhstan on Execution and the Law on the State System and "On 6paque and Family".

Thus, the Law on marriage and family states that the age of marriage for men and women is 18 years. Article 9 of the law requires that both men and women have entered into marriage before the age of 6pa6o and have reached the age of 6pa6o. Article 29 of the law states that spouses have equal rights and equal obligations. Each spouse is free to choose their occupation, profession and place of residence. Issues such as childbirth, parenthood, upbringing and education are decided by husband and wife. Husbands and wives are expected to establish family relationships on the basis of mutual respect and mutual assistance, in order to strengthen family health and well-being and to take care of the health, growth and well-being of their children.Since then, many laws have included provisions that actively discriminate against members of the opposite sex.In particular, the Penal Code (the "Penal Code of the Republic of Kazakhstan"), the Code of Penal Procedure (the "Code of Penal Procedure of the Republic of Kazakhstan") and the Code of Penal Execution (the "Penal Code") are referred to as the "Penal Code".) The Labour Law of the Republic of Kazakhstan (hereinafter referred to as the "HCK PK") does not provide for gender discrimination in the Labour Law of the Republic of Kazakhstan (hereinafter referred to as the "Labour Law"). The Law on Administrative Offences of the Republic of Kazakhstan, the Law on Pensions of the Republic of Kazakhstan, etc.For example, under the rules of Article 42 of the Criminal Code of the Republic of Kazakhstan, women over fifty-five and men over sixty years of age, pregnant women and women with children under three years of age are not entitled to perform such duties.Detention does not apply to women and the Law on the Prevention of Domestic Violence establishes the legal, economic, social and organizational framework for state and local government institutions,organizations and citizens of Kazakhstan to prevent domestic violence. Bce laws are aimed at protecting the constitutional rights, cu6od and legal rights of man and citizen in o6lacty of family and family relations, preventing and stopping domestic violence, identifying and eliminating the causes and o6ctoiatects that benefit them. Committee. The adopted laws created oco6cient legal conditions for the prevention and termination of domestic violence against women.The Law on Gender Equality aims to regulate public relations in the area of national guarantees, equal rights and equal opportunities

for men and women, and establishes the basic principles and norms of creating conditions for gender equality in all spheres of national life and public life [25].
At the same time, an analysis of existing national legislation, which ensures equal rights and opportunities for men and women in all spheres of life in the country and in society, shows that in terms of the law, Kazakhstan has created a principled environment. Established to maintain gender equality. Kazakhstan has acceded to and ratified almost all of the gender equality conventions of the United Nations. However, despite the existence of legal equality, there are some differences between the reality of formal equality and equal opportunities.
Thus, after the OOH Committee on the Elimination of Discrimination against Women reviewed the second periodic report of Kazakhstan at the 757th and 758th sessions of the 37th Conference of the OOH on 16 January 2007, it delivered its final statement on the gender situation in Kazakhstan [26]. In particular, the Committee is concerned that the provisions of the Convention, the Optional Protocol and the Committee's recommendations, including among judges, have not been widely disseminated.
The Committee calls upon States parties to take active measures to disseminate the Convention, the procedures under the Optional Protocol and the Committee's recommendations and to ensure that prosecutors, judges, members of the judiciary and lawyers carry out programmes covering all aspects of the Convention. Additional agreement.
The Committee also called on Kazakhstan to ensure that public officials, in particular law enforcement officials and judges, fully understand existing legislation on domestic violence, understand the different forms of violence against women and take appropriate action.
In this connection, Kazakhstan adopted an Action Plan to implement the recommendations of the 37th session of the United Nations General Assembly on the Elimination of Discrimination against Women, which was adopted following consideration of the second periodic report of Kazakhstan. Implementation of the Convention on the Elimination of All Forms of Discrimination against Women [27].

In order to implement the above-mentioned action plan, a manual for judges has been prepared, which includes United Nations legal instruments on equal rights for men and women, as well as Kazakhstan's current legislation and provisions on gender equality.In accordance with Article 4 of the Constitution of the Republic of Kazakhstan, international treaties ratified by the Republic of Kazakhstan take precedence over its laws and are directly applicable.In connection with the above, the annex contains other documents relating to the international treaty law of Kazakhstan and the Republic of Kazakhstan and their

application in the national legislation of women with minor children (article 46 of the Penal Code).
In Kyrgyzstan, there is a relatively strong legislative system that guarantees equality between men and women. Thus, civil, criminal, labour and family laws all declare equal rights for men and women. In 2013, the Office of the President of Kyrgyzstan approved the Roadmap for Sustainable Development for 2013-2017. These measures will support the implementation of the first long-term innovative strategy on gender equality (2012-2020) and the National Action Plan in early 2012.
In February 2013, Kyrgyzstan, in cooperation with the OCHA Center for Preventive Diplomacy in Central Asia (CPCAD) and the Organization for Security and Cooperation in Europe (OSCE), adopted a National Action Plan with the active support of the OOG. 25] Together with OOG Security Council Resolution 1325 and OOG Security Council Resolutions 1820, 1888, 1889, 1960, 2106, 2122, it is part of OOG's commitment to women's rights. This policy document represents a new approach to gender equality in the Kyrgyz Republic.
Despite the measures taken, the country still has high levels of inequality and large disparities between regions. Women are virtually excluded from decision-making processes. Violence against women is widespread and takes many forms, including domestic violence, bride kidnapping, trafficking, early marriage and physical violence. Misinterpretations of certain cultural and social practices increasingly limit women's right to self-determination. This increases the risk of women becoming involved in radicalized religious groups.
Unpaid home-based work limits employment opportunities for rural women in agriculture or other professions. Women and girls in the Kyrgyz region have limited access to productive resources.

The first full-fledged OOH Office for Women in Central Asia in Kyrgyzstan, which has been operating since February 2012, actively supports the rights, empowerment, equality and dignity of women and girls in the country.
The Government of Tajikistan has also developed a good system for equalizing the rights of men and women, as well as implementing all the points of the Beijing Declaration, also signed by Tajikistan in 1995. Since then, four important objectives in the area of gender equality have been achieved. [27] These include: 1) establishing a legal framework on this issue; 2) writing strategies that prioritize gender equality issues and suggestions for solutions; 3) establishing presidential quotas for the recruitment and creation of women in rural and mountainous areas; 4) establishing principles of equality in the areas of land and housing at the legislative and policy level.

Tajikistan, after the collapse of the Soviet Union in 1991, obtained the status of a presidential republic, faced a number of problems, which included a civil war that lasted on the territory of Tajikistan for about five years, thereby slowing down the level of development of the state.
After the end of the war, Tajikistan set the goal of establishing a democratic state, liberalizing the economy and building a civil society. Progress could not be achieved without introducing conditions for women and increasing their role in all spheres of activity. Thus, in 1994, in Tajikistan, a national referendum enshrined the official Constitution, which included a clause on the equality of citizens, regardless of sex, nationality, age, religion, property and other similar factors. This article is responsible for equality between men and women in all respects. Also, the Tajik Constitution states that families are protected and supported by the state and is the basis of their o6presentation. For example, in case of divorce between wife and husband, they have the same rights and privileges. The prohibition in the Constitution is multitudinous. Article 34 of the Constitution states that the mother and child are subject to the protection of the state. There are other clauses in the Constitution that give a guarantee of equal rights for women. These include articles such as: Article 27 on equal guarantees, irrespective of sex, for all Tajik citizens to participate in the political and public life of the state, Article 32 on equal rights and guarantees of inheritance, Article 35 on guarantees of equal pay, Article 38 on guarantees of health care, Article 39 on guarantees of social protection, Article 41 on the right to higher education. All of these articles have become a fundamental basis for the further development of other projects that promote equality in Tajikistan.
In the late 1990s, the Government of Tajikistan issued a National Strategy of Action to Increase the Role and Status of Women for 1998-2005. [28]It included such tasks as conducting gender statistics on legislation, increasing women's role in such areas as economics, politics, finance, education and others, adopting a number of programs on this topic, programs to protect women's health and so on. The Government adopted a State programme entitled "Guidelines for State policy to ensure equal rights and opportunities for men and women in Tajikistan for the period 2001-2010", which guarantees not only an improved status for women and guarantees equal rights for them, but also mechanisms for preventing violence against women. The implementation of all the above-mentioned goals is planned to be achieved through institutional mechanisms.
In 2002, the Government of Tajikistan introduced the Presidential Quota for 2001-2005, which grants free higher education to young girls graduating from isolated and rural areas, as well as from poor families and those with many children. A few years later, the quota was expanded to include not only girls but

also boys, i.e. high school graduates. In the following years, this quota was reproduced again, due to its efficiency and indispensability. According to the Tajikistan Government's official statistics, in 2011, 595 seats were allocated, in 2012 - 617 seats, in 2013 - 636 seats, in 2014 - 652 seats and in 2015 - 676 seats. In 2005, a new Law of Tajikistan "On State Guarantees of Equal Rights and Opportunities for Men and Women" was issued. For the first time in Tajikistan's legislation, the term "gender" was introduced and further explained, as well as similar
"gender policy", "gender equality" and other terms. Article 1 of the Law provides a detailed explanation of the term, while Article 3 states the prohibition of discrimination on the basis of sex. The law also guarantees punishability of all actions performed against the law and the above-mentioned principles of gender equality [29].
A programme plan entitled "Upbringing, training and placement of qualified women and girls in leadership positions in Tajikistan for the period 2007-2016" was also adopted for the purpose of identifying and promoting women. According to this document, the objectives pursued by the Government of Tajikistan include: 1) o6ecopoea ling all proper conditions for o6pa6oing women; 2) building o6effective housing for female scholarship holders, female students; 3) agitating for girls to enter 10-11school grades; 4) equal privileges for women and men in leadership positions, and so on.The above-mentioned official documents of Tajikistan are a direct proof of the fact that this country is committed to the principles of the Beijing Declaration and also to the legal basis for the improvement of gender equality in society.Since the first days of its independence, Poland has demonstrated to the entire world that it adheres to its democratic principles by acceding to some 70 major international human rights instruments. These include the Universal Declaration of Human Rights, the United Nations Convention on the Elimination of All Forms of Discrimination against Women, the Millennium Development Goals and others.Building on the co6cutionary model of socio-economic planning, pecy6pa6lical plays a key role in addressing national development issues through civil society development, gender equality and women's empowerment in the o6ciety and the state, institutions and private sector as important partners. The development and implementation of economic and legal mechanisms.The equality of rights of women and men is enshrined in the Constitution of Pecpy6 icatan, Article 18 of which states: "All citizens of Pecpy6 icatan have equal rights and rights and are equal before them regardless of sex, race, nationality, language, religion, social origin or y6e origin, law personal and social status [29].
The Constitution of Peculiara Yoz6ekitana grants everyone the full range of

personal, social, political, cultural and economic rights contained in the International Bill of Human Rights. The Basic Law guarantees everyone the right to life, cov6odity and security of person.

Over the years of independence, the country has adopted more than a hundred legislative acts concerning the rights and rights of citizens. They are: "On elections to the Oliy Majlis; on access to court in cases and decisions that violate the rights of citizens; on social protection of persons with disabilities; on civil liability; on the rights of citizens, "O6paozoanization", "On guarantees of iz6paotelnyh rights of citizens", "O6paopaschestvo citizens", "On the cvoo6od of access to information", "O6privatelnosti o6edinenii", "On employment", etc.

Equal standards for the full participation of women in the election process and of men in the election of men in the country are laid down in the Oliy Majlis and Oliy Majlis Elections Acts. The Law of the Republic of Tajikistan "On the Rights of Citizens to the Councils of People's Deputies, Districts and Cities", "On Guarantees of Citizens' Rights to Elections", "On the Enactment of Bills", "On Referendum of Peculiar Republic of Tajikistan", "On Elections of the President of Peculiar Republic of Tajikistan". Pecpy6lika Yoz6eKicTaN ". Article 5 states: "Direct or indirect restrictionThe right of citizens to exercise such rights is not allowed, with the exception of the requirements established by the legislation on expulsion.

The Oliy Majlis Elections Act of 29 August 2003 (art. 22) provides that at least 30 per cent of the total number of candidates put forward by a political party must be women. The Law on Guarantees of Citizens' Rights guarantees the protection of citizens' rights regardless of their gender (Article 20).

The Family Code of Pecpy6ylica Yoz6ektitana (1998) provides for equality of personal and property rights of married men and women (Article 2) and for equality of spouses in child-rearing and care, as well as in all other family matters. (Article 21).[30]

Recognizing that women's and girls' access to education, vocational training and the development of their competencies is key to their empowerment and advancement, the country guarantees the right to education to all citizens, regardless of gender. Yz6eqiqtan (Article 41). Equal rights of women and men to exercise are also enshrined in the Law on Physical Education and Sports (Article 2).

For 6o greater representation of women in the decision-making process, the President of Pecpy6lika Yoz6eqitana adopted the Decree of 2 March 1995 "On enhancing the role of women in the state and o6public structure of Pecpy6lika Yoz6eqitana", "Additional measures to support. Women's Committee of Yz6eqitan "May 24, 2004. Resolutions of the Ka6einth of Ministers are adopted.

The Labour Code of Pecpy6 icc ation (1995) provides for a protectionist approach to women's employment. A pa6odont does not have the right to refuse a pa6odont due to the 6emergency of a woman or the 6o ness of her children. The articles of the Code contain provisions to protect women from working in harmful or difficult conditions.

2.2. Leading Women Leaders in Central Asia: Country Analysis

In Peculiar Kazakhstan, men and women have very different ideas about governing capabilities, as well as different political views and ways of developing the state. Women are critically under-represented in various political structures of power, which means that women cannot form a "critical minority", i.e. about 30-40% of the time their voice was "heard" and had an impact on the political situations in country. Thus, at the beginning of the twentieth century, Kazakhstan achieved a high level of performance in the field of education, so the government set new priorities as strengthening the role of women in the political life of the society, as well as in the spheres of legislation and executive power. According to statistics from the Kazakhstan Civil Service Agency, the percentage of women in leadership positions has risen to 52% of the total workforce, but only about 10
% of women hold high positions in political bodies.

In 1991, Kazakhstan became independent, after which there were some changes in local government: the number of women in national and local government sharply decreased. Twenty years later, after the elections at the beginning of 2021, the number of women in the lower house of Parliament, i.e. in Maslikhat, has increased from 16% to 21%, i.e. by six people, while in the upper house of Parliament, namely in the Senate, the situation has not changed. Women are about 5% of the total number of assessors.[31]

Everywhere else in Kazakhstan, the number of women in Maslikhat is many times lower than that of men: only 20% of the o6 total number of deputies. The Maslikhat region plays a very important role. Thus, in northern Kazakhstan, the percentage of women in the lower chambers is higher than in southern Kazakhstan. For example, in Pavlodar Region, the percentage of women in the Maslikhat is around 20 per cent, while in South Kazakhstan Region the percentage has never been higher than 5 per cent of the total number of assessors. In the southern region, the number of women is decreasing every year, despite the fact that every year more women are nominated as candidates for election.In the upper house of the Senate, there are 48 seats, of which only two are occupied, i.e. only 3.9%, according to the Parliament of the Republic of Kazakhstan in 2010. In the lower chamber, Majilis, the number of seats is 110, of which 21 seats are held by women, which is about 24% of the total number of deputies.Based on all the above data, it is worth concluding that women are critically under-represented in the government, and male representatives are clearly dominant. In fact, women account for most of the total number of women in the government, but there are more men in the central executive branch, and women account for only about 9% of the total. Of the 18 ministers in Kazakhstan, only three are women, namely the Minister of Economic Integration, Labour and Social Protection. There are also seventeen line ministers and one prime minister. In total, there are about three executive secretaries in each ministry and about four vice-ministers who are women.Also, in organizations such as local and regional governments, it is very rare for a woman to hold one of the leadership positions. Women usually occupy the lowest positions.Oco6eno such cases are peculiar to rural areas. Of all 17 o6lac¬tions in Kazakhstan, there is only one woman akim - Gulshara Naushaevna A6dikalikova, who has held the position of akim of Kyzylorda o6lac¬ti since the end of March 2020. In Kazakhstan, only five women have ever held the position of deputy akims. At the moment, only three women serve as rayon akims at the rayon level, 20% of the total number of rayon deputy akims are women. Among women leaders in rural areas (this also includes villages, towns, regions, etc.) a total of 262 women akims, or 9 per cent of the total. [33] These figures demonstrate that power is not a pacpopular phenomenon among women, and the voice of women in these bodies is not as important as that of men, because it is the dominant one. The presence of women in local government is very important for building a democratic state.From all of the above, the data shows that only 33% of women are represented in power structures. At the moment, there are no mechanisms in Kazakhstan to correct these shortcomings, so there is very little opportunity for change in the near future. Kazakhstan, as a party to the Convention on the Elimination of All Forms of Discrimination, is obliged to provide equal opportunities for all genders. The Government has already planned to address these inequalities by issuing a quota for all those who 6are elected to Parliament in order to get rid of gender inequalities, but this plan has not been successful. [34] In recent years, there has been a clear increase in women's activism and interest in public-political co6 stances. For example, the number of women who participated in the most recent parliamentary elections almost doubled, but the final composition of the deputies did not change in any way. None of the currently active political parties in Kazakhstan have focused on the concept of gender equality in their programs. Therefore, no effective strategies have been developed to involve women in political change.

Of the other measures provided by the Government of Pecuny6lika Kazakhstan are the Women's Leadership Capacity Building Schools, sponsored by regional organizations. These schools provide the full spectrum for women's development in all spheres. Non-governmental organizations and international organizations also contribute the most to development. Among the most effective ones is the "Tomaris" programme, which operated for three years. This programme resulted in the creation of Women's Leadership Schools. These schools represent a network of NGOs, currently there are 60 of them in Kazakhstan, and there are also subsidiary schools for women politicians in Hyp-Sultan and Almaty.According to 2020 data, the following numbers of women occupy leadership positions in Pecy6like Kazakhstan: 3 ministers, 5 executive secretaries, 11 vice-ministers, and one deputy of national committees, and so on.

Other obstacles to the creation of gender equality in the political environment are mentality. The position of women in Kazakhstan's society is not yet something people are used to and does not inspire confidence. For example, a woman candidate is given less financial support for his or her program than a male candidate.A 2009 survey on women's leadership in Kazakhstan revealed that, surprisingly, 70 percent of respondents stated that women have the same leadership potential as men and can be active political participants on an equal basis with the latter. However, since there are many times more men in Kazakhstan's political environment, there is very little help available to improve women's potential.

In Tajikistan, women's participation in decision-making structures has been consistently low for thirty years since independence. According to the most recent data from 2020, in Tajikistan only 17% of women are deputies in both the Senate and the Majilis in general. In 2014, women's participation in local or central government was only about 21% of the total. In the central executive branch, the results are even lower: only 523 women work in this area, which is 16% of the total number, respectively. However, progress has been made at the local level, as about 38% of women have joined jamoats in the last time [35].

The government of Tajikistan has not established specific quotas for women's political participation, but as in Kazakhstan, there are widespread special schools for capacity building of women leaders, also sponsored by regional organizations. Civil society in Tajikistan has become more active in recent years on gender equality issues, with many non-governmental organizations providing services and developing programs for women seeking to serve in Tajikistan's governing bodies.In Kyrgyzstan, following the loss of sovereignty in the early 1990s, the number of women participating in the political environment decreased significantly. Under the Soviet regime, Kyrgyzstan effectively had a quota allowing 30% of women to be included in various structures (including the Parliament). After the collapse of the Soviet Union, this quota was abolished; subsequently, the number of women in the political environment decreased accordingly. A large number of women are represented at all leels in Kyrgyzstan. For example, in 2002, in the elections to Parliament, out of women MPs accounted for only 5.9% of the total number. At the moment, according to the latest data of 2020, only 2% of the total number of women MPs in the Parliament of Kyrgyzstan are women. Since independence, only the following ministries have had women as ministers: Ministry of Education, Ministry of Justice and Ministry of Labor and Social Protection. In the 1990s, only one woman was appointed Minister of Foreign Affairs, and there were no women at all in the Ministry of Finance.

The above described situation is similar with respect to the district, rural, and o6lacuty levels. According to the National Statistics Committee, in the last 10 years, out of all 260 deputies of local councils, only 11% of the total number were women, which is 27 people.

Recently, most of the electorate in Kyrgyzstan is made up of women, about 63 per cent. In Kyrgyzstan, women have become more active in politics and are an important part of civil society. Despite this, Kyrgyzstan's political parties never raise the issue of gender equality as something important for building a democratic society.[36] The Kyrgyz National Report on Gender Issues notes that women are well aware that men are dominant in politics, but they cannot become equal partners and declare their own ploys and initiatives [37].

It is also important to note that the decline in women's political participation today, compared to the Soviet period, can be explained by the following: stereotypes embedded in traditional Kyrgyz society have not been eradicated, as was planned in the Soviet Union; rather, these nuances were skilfully concealed.[38] International organizations, while studying Kyrgyzstan's political o6cation, have concluded that not only is there a lack of resources and opportunities for equal participation in politics, but also the fear of women that the o6ciety is not accepted to engage in traditionally "unladylike" activities.

Turkmenistan, like the other countries of Central Asia, has signed all the relevant conventions and instruments related to improving equality within the state. These include the Convention on the Elimination of All Forms of Discrimination against Women, which they ratified in 1997, and the Optional Protocol to the Convention, to which Turkmenistan acceded in 2009. In order to progress, Turkmenistan produces its co6created National Plans and Strategies, which every time address the issue of unequal treatment of women, namely methods for eliminating these problems. The most recent important document was"National Action Programme on Gender Equality in Turkmenistan 2015-2020", which outlined the following Measures and mechanisms to attract women into relevant structures and to increase the employment rate of women in all sectors of production.

An important factor for analysis is the degree of women's participation in the political processes of the state. According to the latest information, there have never been any quotas that would help women to be active in politics in Turkmenistan. According to the results of the fifth parliamentary elections in 2013, the number of women MPs in the Turkmen Parliament was about 27%, which exceeds the international norm of only 22%. In Turkmenistan, there have also been cases where a woman has held high positions in the Mejlis (e.g., vice-chairman and so on). [39]With respect to the local, municipal and rayon level, the figures are also quite good. At the moment, women's participation at the city and district level is about 20-22%, at the local level about 16% and at the rural level about 18% respectively.Uzbekistan is a country where there have been positive developments that have resulted in an improvement in the status of women in the country compared to the experience of other countries. For example, from independence in 1992 until the early 2000s, the participation of women in the political sphere was high, with women representing 9.5 per cent of the population. In 2003, a quota system was introduced by the government in relation to the Law on Execution. These quotas allowed for the participation of at least 30% of women in the political sector of the state. It was this innovation that helped Yezekaterinburg to achieve sufficient gender balance in the political system. Prior to the introduction of constitutional reform and the creation of a bicameral Parliament, the number of women in the Parliament was less than 7%. After the elections in 2005, they were given new seats with 17%.[40] Despite this, there are relatively few women in high-level positions in the State. A critical minority is thought to be between 30-40 per cent, while in Yz6ekistan women account for only 15 per cent of the Senate and 20 per cent of the members of the Legislative Chamber, a mere 19 per cent of Oliy Majlis.

There are 11 parliamentary committees, none of which has a female chairperson. Only the committees dealing with water, agriculture, agriculture, construction and commerce are women [41].

All of the political parties in Yuz6ekistan have a women's wing, but none of them promote gender equality or the other issues related to it as something important. It is important to mention that since the introduction of the quota described above, there have also been positive changes in political parties: the number of women has increased from 37% to 50% in the top five parties.[42] However, there is a ppo6lma about the quantitative aims of this quota, so women are notare so active in the process of functioning of parties and do not occupy leading positions in them.Women are particularly active in the work of non-governmental organizations, of which there are 210 throughout the country. These NGOs include both state and local, as well as special women's units on specific important issues. At present, of the total number of about 63 NGOs, there are about 6a6. [43]

2.3. The role of international organizations in the development of women's civic engagement

Since the beginning of the 1990s, following the loss of sovereignty, various international organizations and international financial institutions have been active in Central Asia, not only assisting and supporting states, but also conducting analyses of the gender situation in each country. UNICEF, the Asian Development Bank, the World Bank, the structure "OOH-Women" and others are considered to be among the most active on this issue. They develop strategies that aim to support women in such areas as entrepreneurship, politics and the eradication of domestic violence and all forms of discrimination against women.Recent trends in the aftermath of the coronavirus pandemic have had a major impact on the success of policies pursued by these organizations. Due to socio-economic crises in States, social minorities have been the hardest hit: women are also in this category. At this point, recovery from these crises must take into account gender policies as well as improved protection of women's rights.

The Asian Development Bank has been an active gender champion in the Central Asian region. ADB's gender strategy was launched in 1998 and includes the integration of these issues as one of its key themes. An important part of ADB's mandate is to write and analyze reports on this topic, covering such areas as the economy, social welfare, protection of women's rights, employment, and the future that Central Asian countries should pursue. In its analysis, ADB has mainly focused on various obstacles to achieving gender equality in all spheres of social life [44].

"There is a critical need for additional efforts to ensure that women's voices are considered at all stages of planning and implementation of regional projects," said Vice President of the Asian Development Bank (ADB) Shixin Chen, co-chair of the conference.

"Gender equality is essential for sustainable development, and a gender strategy will cnoco6ctvye empower women by providing access to capacity building, economic opportunities, and participation in decision-making" [45].

Central Asian countries face a number of significant challenges in improving women's lives. There is a systemic gender gap in labor force participation, with disparities in employment rates, wages, and quality of employment. There are also challenges in implementing national policies on flexible work arrangements, maternity leave and equal pay.

In many countries, women's entrepreneurship suffers from a lack of funding for start-ups and business expansion due to women's limited ownership of assets for credit, while the proportion of women in leadership positions remains low. Women are also often hardest hit by external shocks, such as droughts and floods related to climate change, food and oil price volatility, and major pandemics, due to their disproportionate exposure to risks and family responsibilities. ADB's gender strategy for the Central Asia region describes how the program's activities will complement national strategies for women's empowerment. This includes promoting economic opportunities for women in formal wage employment, in agriculture and in the informal sector, and in entrepreneurship, among others. The Strategy identifies how interventions will address the specific needs of women and girls through project consultative processes. It also seeks to ensure that women have equal access to quality education, health services, and information and communication technologies. "ADB views gender equality as both a critical end goal and a key driver of sustainable social and economic development. The gender strategy adopted today is a step toward closing the gender gap that is creating unhealthy conditions, limiting opportunities, and undermining the empowerment of women and other vulnerable groups in all CAREC countries; it will help ensure that CAREC activities leave no one behind" [46]. ADB notes in its annual reports that civil society development is an important condition for gender policy development in the region. ADB has recommended that the five Central Asian countries need to educate women about their potential for political participation. It is also important to mention that it is necessary to increase human resource capacity, especially of women, to realize gender equality, i.e. these issues will be addressed much more effectivelymore effective and efficient. For education purposes, the Asian Development Bank has offered its gender-related investment initiatives to four Central Asian countries. These are Kazakhstan, Kyrgyzstan, Uzbekistan and Tajikistan. For the important issue of domestic violence, ADB has also developed principles and recommendations to eradicate it. These include: establishing more crisis centers, crisis lines, shelters and women's shelters. These organisations should offer a range of services so that women are not forced by economic dependence to return to dangerous conditions with a tyrant. Services should include financial, legal, medical and psychological counselling as well as opportunities for skills development and income generation. ADB has also noted that it is important to engage more external treatment specialists to deliver trauma counselling that is not clinical in nature and therefore does not carry a social stigma. Such treatment can be provided through non-governmental organizations and other entry points to coo6ect to address the ongoing post-traumatic effects of the civil war and violence among men and women. A very important step is the establishment of residential alcohol and drug treatment programs for men and violent offenders who have addiction problems, taking the violence outside their homes, as well as providing trauma counselling and skills development. To prevent the above challenges, awareness of crime and violence against women and women's rights to protection under the law and to teach boys to respect girls at an early age must also be included in school curricula, beginning in primary school. To do this, the government must provide adequate funding and technical support to develop culturally appropriate advocacy, education and media campaigns that challenge pervasive social attitudes and claim that "culture" and "family" traditions condone violence against women; Implement an awareness-raising programme for police and judicial authorities to help them recognize that domestic violence is a criminal offence and not a private family matter; Support widespread capacity building to ensure that the police and law enforcement system can apply the new legislation; provide specialized training integrated into training programmes for all health professionals to identify and address the physical and psychological effects of domestic violence. Medical pa6o6o nts also need o6y training and should testify in court as new laws come into force. The right skills are also needed for people who are with victims of sexual violence. All Central Asian countries have adopted the Sustainable Development Goals and key international agreements on gender equality, including the Convention on the Elimination of All Forms of Discrimination against Women. The Gender Strategy of Central Asia includes support for policy reforms that address existing gender gaps."OOH Women is the United Nations Entity for Gender Equality and the Empowerment of Women. The organization was founded to accelerate progress for women and girls around the world in order to meet their needs and needs. OOH Women supports its Member States in their efforts to meet global standards for achieving gender equality, and works with national governments and civil society to develop laws, policies, programmes and services that effectively implement these norms and achieve real benefits for women and girls around the world. The organization works to translate the vision of the Sustainable Development Goals into reality for women and girls and to promote equal participation of women in all aspects of life, focusing on five priority areas Expanding women's leadership and political participation, eliminating violence against women, engaging women in all aspects of peace and 6e6oocnocy, empowering women economically, and placing gender equality at the forefront of national planning and 6e6oocnocy. The OOG Women also coordinates and supports the entire OOG system to achieve gender equality, as well as all discussions and agreements related to the 2030 Agenda. The Entity works to define the central role of gender equality in the achievement of the Sustainable Development Goals and inclusive peace.

Since independence, support for women at all levels has been on the public policy agenda. Kazakhstan was the first Central Asian country to establish a national body to promote gender equality in all spheres of life

- The National Commission for Women's Affairs and Family and Demographic Policy under the President of Kazakhstan.

The main legislative act in the field of gender policy is the Law "On State Guarantees of Equal Rights and Equal Opportunities for Men and Women" adopted in 2009, a number of specific indicators to achieve gender equality in politics, economy, education, family, health care and prevention of violence against women and children. In 2016, President HaZap6aeйapproved the Concept of Family and Gender Policy until 2030, the implementation of which is designed to ensure equal enjoyment of all rights regardless of sex and to prevent discrimination and gender asymmetry.

Each of the Central Asian countries has ratified a number of fundamental international instruments, including the OOH Convention on the Elimination of All Forms of Discrimination against Women (CEDAW), the Beijing Declaration and Beijing Platform for Action, the Convention on Women's Political Rights, the Convention on the Nationality of Married Women, the six International Labour Organization (ILO) conventions and the 2030 Agenda for Sustainable Development.

A study supported by OOH Women for the first time in Central Asia found that 17% of women aged 18-75 who were ever in a relationship had experienced physical or sexual violence by a partner, and 21% had experienced psychological violence [47].

In addition, OOH Women in Central Asia oversees the OOH Interagency Thematic Group on Gender, conducts active outreach to raise awareness of gender equality issues, and collaborates with the private sector to promote women's economic opportunities [48].

CONCLUSION

Because gender is considered to be the primary status or primary characteristic by which people identify with them, 'women' are considered to be a 'social minority' in politics. In other words, 'women' must have certain political priorities (o6pecially those related to children and o6pao nition) that o6each women as a distinct 6o ry or group of individuals who tend to have o6political views and preferences.The gender divide in political and public life in the Central Asian region that is now evident has not been a common phenomenon in other periods.

B The ancient Turkic o6ectvaxes that previously lived in the Central Asian region, the woman was a full-fledged member of the o6ectvah. The main activity was not limited to procreation, child-rearing and housekeeping. Turkic women, along with men, were active in military enterprises and were trained in the art of war, as well as participated in military campaigns. Before the introduction of Islam, women had a voice in events such as kurultais, and rulers in the Turkic tribes conferred with their spouses during important decisions.

Despite the fact that women in the traditional Turkic society were in most cases responsible for the household, they had a rather high position in the collective. They had the same privileges and rights as men in relation to family and 6paqy. In general, the high status of women was connected with their increased role in the economy of the state, as women were mainly engaged in economic activities. The disintegration of the Soviet Union initiated 6ecpe6 tionary socio-economic changes in the successor countries and changed the gender 6a lane in the region, considering gender equality as one of the most important legacies of the socialist past. Transition experiences in the region have demonstrated that changes in gender 6aalance caused by economic shifts are rather slow, and that economic growth and women's economic empowerment do not always go hand in hand. Strong measures to empower women economically should therefore be central to the Central Asian policy dialogue aimed at reducing poverty and inequality and inclusive growth.

C The Soviet Union's early leaders were committed to the ideal of "empowering" women in all regions of their country

- The ideal of placing women on an equal footing with men in all aspects of economic, social and political life, which simultaneously providing them with full material and moral support in their role as mothers. However, in different regions of the country, this goal was approached in very different conditions, for different reasons and in different ways. As noted earlier, women's emancipation was primarily a response to demographic, economic and revolutionary challenges; the idea of female emancipation itself was the result of attitudinal and social processes already taking place between the sexes. Soviet policy was mainly "legalistic", which for the most part implied changes in family and labour legislation.In the five peculiarities of Central Asia, Kazakhstan, Uzbekistan, Turkmenistan, Tajikistan and Kyrgyzstan, on the other hand, women's emancipation marked a dramatic shift away from social and economic norms to religious and traditional practices. Women were shut out and excluded from all aspects of social, economic and political life; they were traditionally tied to domestic tasks and childbearing. In general, women were seen in their own right and by others as the lowest of the low. The very idea of women's inclusion was imposed from above by the rebel government as a means of bringing about certain social and revolutionary changes, not caused by economic, demographic or social problems emanating from the society itself. Soviet policy approaches shifted from coercion to support for the population, to an emphasis on social engineering, marked everywhere by bloody and violent reactions from both local men and women.Despite these differences, many Soviet writers now claim that full gender equality 6 was achieved in all parts of the CCCP. Soviet emancipation policies allowed and even forced women to participate in the industrialized economy. They became factory workers, tractor drivers, heads of collective farms and farms. The rest became members of local and all-Union vy6opnyy soviets. In 1930, the Soviet government began a campaign for a universal elementary education. Within ten years, enrollment in schools in the Soviet Union tripled, and about one third of the enrollment was in non-Russian pec6y6 lics. This upsurge has put forth a major initiative to recruit and o6ypecialize teachers locally. Overall, the number of teachers in the Soviet Union increased from less thanWomen played a crucial role in filling this niche, from 400,000 in 1928 to 6 million by 1939. Throughout the CCCP, not least in Central Asia, teaching would become a predominantly female profession. In the capitals of Central Asia, institutions of higher learning opened, and women graduates often pursued careers in science. In Soviet society, Central Asian women were at the same time seen as symbols of women's "ocupo6o nition".The collapse of the Soviet Union marked the beginning of a social and economic transformationCentral Asian economies. Their experience has shown that the changes in gender induced by economic shifts are far from obvious. While Central Asian women made significant economic and social gains during the Soviet years, these gains have in many cases been reversed in the post-Soviet period, partly due to the economic shifts experienced by countries in the region, including private sector growth and significant migration flows, and to some extent due to the strengthening of patriarchal traditions.The post-Soviet countries of Central Asia experienced a 6certain transition to a market economy in the 1990s. Many inhabitants of these countries witnessed a sharp decline in their quality of life, but some others were able to benefit from the opportunities presented to them by the transition to capitalism, thereby advancing their economic 6alldominance and advancement on the social ladder. However, this transition was definitely a gender phenomenon: the gender dynamics of opportunity and social mo6ilocity led to different experiences of men and women. Data show that women are lagging behind men in employment outcomes, according to the World 6aank's 2021 results. Despite continuing social and political changes, Central Asia's societies are overwhelmingly patriarchal, and women are grossly underrepresented in positions of power and influence in government. Family is also prioritized over o6pa6cation and career goals for 6o majority of women. Despite the fact that 6o majority of women in Central Asia obtain higher education, their experience in the labor market is very different from that of men. A large percentage of women are employed in the informal sector, with long working hours and unfriendly policies towards family workers. In the business sector, women lack the skills and capital required for productive entrepreneurship Although legislation protects women from gender discrimination, women continue to face gender discrimination in politics, economics and entrepreneurship. To date, women in Central Asia have been frustrated by the lack of opportunities and have been reluctant to acquire the necessary connections and capital to find employment, particularly in managerial positions.Based on the above judgments, improving gender policy - This is key to the development of a society, democracy and sustainability in an era of globalization. In order to achieve gender balance in all branches of government, the government should focus on gender statistics and develop democratic institutions within the state that are responsible for gender equality.The following measures are necessary to achieve gender equality in Central Asia: Institutionalize equality measures to the maximum extent possible to induce structural change; adopt clear guidelines that explicitly establish gender equality as a criterion for nomination and selection; introduce formal, open, transparent procedures for candidate o6opation and nomination; this includes making gender equality the goal of the electoral process. In Central Asian states, the practice of nominating a list of candidates for particular positions should be gender-sensitive is not uncommon. When nominating candidates for international positions, Member States should consider nominating at least two people - one woman and one man (where this is consistent with the constituency process). When voting for these positions, countries should consider gender parity as a goal.

LIST OF REFERENCES USED

1 Shegenova.GenderPolicyofPecuny6likaKazakhstan,2012 https://cyberleninka.ru/article/n/gendernaya-politika-respubliki-kazahstan

2 Beke6aeva A.D. Policy towards women and men in modern Kazakhstan, 2016, p.6-12.

3 Burkhanov K.H., Baisakova Z.M., Ishmukhamedova L.I., Kaltaeva L.M., Ko6eeva A.O., Shakirova S.M. Women's movement of Kazakhstan as an aspect of social modernization, 2011, p.45-46.

4 Lipovka A. Gender Segregation of Labour in Higher Education in Kazakhstan,
2018, c.8-11.

5 Dluzhnevskaya, G. B. 2000. The complex of the Old Turkic time in the cemetery Ylug-Byuk-II. B: Molodin, B. I. (ed.), Monuments of the ancient Turkic culture in the Sayan-Altai and Central Asia. Hovocy6ipcq: NGYY, pp. 178-188.

6 Duman, L. I. 1970. The foreign political relations of China with the Hsiung-nu in the I-III centuries. B: Tikhvinsky, S. L., Perelomov, L. B. (eds.), China and its neighbors in antiquity and the Middle Ages. M.: Nauka, p. 37-50.

7 Early Turks: Essays on History and Ideology. Almaty: Daik-Press, 2003 p 110.

8 History of Central Asia and monuments of runic writing, 2001 Cp6.: Cp6Y, p. 78.

9 Tang state tactics in 6op6e for hegemony in eastern Central Asia. B: Derevyanko, E. I. (ed.), The Far East and Neighboring Territories in the Middle Ages. Hoovocio6ipcк: Nauka, pp. 103-126.

10 B. B. Tishin, H. H. Seregin. Women in the Old Turkic Society, p. 125

11 Seregin, H. H. 2013. Social Organization of the Early Medieval Turks of the Altai-Sayan Region and Central Asia (Based on the Materials of the Pogpe6aal Complexes), p.123.

12 Social history of the Scythians. The basic concepts of the development of the ancient nomads of the Eurasian steppes. M.: Nauka. 1975, p. 89-90.

13 B. B. Tishin, H. H. Seregin. Women in the Old Turkic Society 127

14 Aivazova S. G. Gender equality in the context of human rights. M.: Eslan, 2001.

15 WorldBank,ejergoodoachievement https://www. worldbank.org/en/country/kazakhstan/overview

16 Ministry of National Economy of Kazakhstan. Committee on Statistics. "Women and Men in Kazakhstan in 2012-2016", Astana, 2017.

17 Tokaev, K.K. Foreign policy of Kazakhstan in the conditions of globalization / K.K. Tokaev. - A., 2000. p.3.
18 Asian Development Bank (2013), Kazakhstan:Gender Country Assessment, Philippines,www.adb.org/sites/default/files/institutional-document/34051/files/kazakhstan-country-genderassessment.pdf.
19 European Commission (n.d.), "A Guide to Gender Impact Assessment", Directorate-General for Employment, Social Affairs and Equal Opportunities, EuropeanCommission,Belgium,
http://ec.europa.eu/social/BlobServlet?docId=4376&langId=en.
20 Ombudsman of the Republic of Tajikistan. 2011r. The strategy for 2011-2015. Dyashan6e. Committee on Women and Family Affairs. p. 56, 2014r. ,
21 Press conference of the Committee on Women and Family Affairs under the Government of Tajikistan. 30 July. http://kumitaizanon.tj/ru/news/id/150.
22 A6dypaZakoVa D. and J. Gardner. 2012 г. Gender Statistics in Southern Transcaucasia and Central and West Asia: A Situation Analysis. Manila: Asian 6aank Development. Ak6apoйa F. 2010.
23 Women's economic empowerment. Presentation at
43rd Annual Meeting of the ADB Board of Governors, Tashkent.
24 The State Statistics Committee of Turkmenistan. About Turkmenistan. (available at: http://www.stat.gov. tm/ru/main/info/turkmenistan/).
25 World Bank. O6zop for Turkmenistan, 2019 7.C.10
26 World Economic Forum. 2014. The Global Gender Gap Report, October 2014.(availableat http://www3.
weforum.org/docs/GGGR14/GGGR_CompleteReport_2014.pdf).
27 Medical and Demographic Survey of the Kyrgyz Pecny6liky 2012.
C. 206 — 207.
28 Program for the Development of Social Protection of the Population of the Kyrgyz Republic for 2015-2017. C.15.
29 National Statistical Committee of the Kyrgyz Republic. 2014. On the Situation of Rural Women in the Kyrgyz Pecy6like. (Available at: http://www.stat.kg/media/publicationarchive/66bd4835-30f0-4dcc-b321-694e9b12e336.pdf).
30 FAO. 2016. National Gender Profile of Agriculture and Rural Households: Kyrgyz Pecpy6lika, p. 64. Country Gender Assessment Series. Budapest.
31 Gender Aspects of Social Policy: Strategies and Levels of Implementation. [Electron.pecypc]. - URL: http://www.gender-cent.ryazan.ru/rabzhaeva.htm
32 Ganieva G. Gender Studies in Central Asia. Gender researches in uzbekistan: modern condition and V international scienti fic conference of the german-kazakh university, Almaty (kazakhstan) 13 - 15 march, 2008 p. 51 – 59)

33 Melanyin M. Gender Aspects of the Social State// o6o zpezpevatel- observer 12/2013, p.30-38). 11 Multi-country Chapter of the Structure of OOH-Women inKazakhstan[Electron.pecypc]. 2015. -URL: http://eca.unwomen.org/ru/where-we-are/kazakhstan-multi-country-office
34 Kandiyoti, Deniz. "The Politics of Gender and the Soviet Paradox: neither colonized, nor modern?" Centralasian survey 26 (4):601-623.),
35 Gender and Nation in Post-Soviet Central Asia: From National Narratives to Women'sPractices[Electron.pecypc2015.URL:
http://www.studfiles.ru/preview/1720050/page:29.(dateofpublication: 15.08.2017)
36 National Statistical Committee of the Kyrgyz Republic. 2013. Kyrgyz Republic Demographic and Health Survey 2012. - National Statistical Committee of the Kyrgyz Republic Bishkek, Kyrgyz Republic Ministry of Health Bishkek, MEASURE DHS ICF International Calverton, Maryland, U.S.A. Kyrgyz Republic. 2013. (Available at: http://dhsprogram.com/pubs/pdf/FR283/FR283.pdf).
37 Zvyagelskaya I. Formation of Central Asian states. Political Processes, p.293. M.2009.
38 Allcott M. A second chance for Central Asia. URL:
http://carnegieendowment.org/files/9429CentralAsia_book-full_text.pdf (accessed 4.08.2020)
39 European Commission (n.d.), "A Guide to Gender Impact Assessment", Directorate-General for Employment, Social Affairs and Equal Opportunities, European Commission ,
Belgium, http://ec.europa.eu/social/BlobServlet?docId=4376&langId=en.
40 OECD (2014a), Kazakhstan: Review of the Central Administration, OECD Publishing, Paris. DOI: http://dx.doi.org/10.1787/9789264224605-en. OECD (2014b), Regulatory Policy in Kazakhstan: Towards Improved Implementation, OECD Publishing, Paris. DOI: http://dx.doi.org/10.1787/9789264214255-en.
41 Constantine, Elizabeth A. 2007. "Practical Consequences of Soviet Policy and Ideology for Gender in Central Asia and Contemporary Reversal." In Everyday Life in Central Asia: Past and Present, edited by Sahadeo, Jeff and Zanca, Russell, 115-126. Bloomington: Indiana University Press.
42 Edgar, Adrienne Lynn. 2006. "Bolshevism, Patriarchy, and the Nation: The Soviet 'Emancipation' of Muslim Women in a Pan-Islamic Perspective." Slavic Review 65 (2): 252-272.
43 Féaux de la Croix, Jeanne. 2013. "How to Build a Better Future? Kyrgyzstani Development Workers and the 'Knowledge Transfer' Strategy." Central Asian Survey 32 (4): 448-461.

44 Finke, Peter, and Sançak, Meltem. 2007. "Konstitutsiya buzildi! Gender Relations in Kazakhstan and Uzbekistan.". In Everyday Life in Central Asia. Past and Present, edited by Sahadeo, Jeff and Zanca, Russell, 160-177. Bloomington: Indiana University Press.
45 Kamp, Marianne. 2009. "Women's Studies and Gender Studies: Are We Talking to One Another?" Central Eurasian Studies Review 8(1): 3-12.
46 Kandiyoti, Deniz. 2007. "The Politics of Gender and the Soviet Paradox: Neither Colonized, Nor Modern? Central Asian Survey 26 (4): 601-623.
47 Kudaibergenova, Diana T. 2016. "Between the State and the Artist: Representations of Femininity and Masculinity in the Formation of Ideas of the Nation in Central Asia". 44 (2): 225–246. Nationalities Papers. doi: 10.1080/00905992.2015.1057559.
48 Massell, Gregory J. 1974. The Surrogate Proletariat. Moslem Women and Revolutionary Strategies in Soviet Central Asia, 1919-1929. Princeton, NJ: Princeton University Press.
49 Roche, Sophie, and Hohmann, Sophie. 2011. "Wedding Rituals and the Struggle Over National Identities." Central Asian Survey 30 (1): 113-128.
50 Tett, Gillian. 1994. "Guardians of the Faith? Gender and Religion in an (ex)Soviet Tajik Village.". In Muslim Women's Choices. Religious Belief and Social Reality, edited by Fawzi El-Solh, Camilla and Mabro, Judy, 128-151. Oxford: Berg.

Printed by Books on Demand GmbH, Norderstedt / Germany